THE NEXT WEB OF 50 BILLION DEVICES
MOBILE INTERNET'S PAST, PRESENT AND FUTURE

Majeed Ahmad

SMARTPHONE CHRONICLE

This publication is designed to provide accurate and authoritative information in regard to the subject matter covered. It is sold with the understanding that the publisher is not engaged in rendering professional services. The advice and strategies contained herein may not be suitable for your situation. You should consult with a professional where appropriate. Neither the publisher nor author shall be liable for any loss of profit or any other commercial damages, including but not limited to special, incidental, consequential, or other damages.

Some of the images on the cover artwork are courtesy of Google, Jolla, Looxcie and Popular Mechanics

Copyright © 2014 Majeed Ahmad Kamran
All rights reserved

ISBN 10: 1499146698
ISBN 13: 978-1499146691
Library of Congress Control Number: 2014907051
CreateSpace Independent Publishing Platform
North Charleston, South Carolina

Dedicated to Professor A. Heetman

CONTENTS

Prologue... 1
Chapter 1: The Next Web............................... 7
Chapter 2: Pre-iPhone Mobile Web 33
Chapter 3: Mobile Internet 2.0 67
Chapter 4: The Internet Versus the Web................ 91
Chapter 5: The Browser Wars 121
Chapter 6: Digital Sixth Sense 149
Chapter 7: Twenty First Century Network 173
Chapter 8: Mobile Commerce: A Case Study 199
Notes..225
Index..237
About The Author...................................243

PROLOGUE

"We've never had the ability in our industry to reach five billion people with a computer and now we have the ability to do that. That's big."
— Marc Andreessen, co-author of the Mosaic browser, Netscape co-founder, and general partner at venture capital firm Andreessen Horowitz

In 2014, when Long Term Evolution (LTE) technology—the manifestation of fourth-generation (4G) wireless systems—was still evolving, the notion of the fifth generation (5G) networks began making waves. While the 5G technology was in an embryonic stage and was a mere concept, the question was "what's the big rush?" First and foremost, mobile Internet traffic was growing exponentially. Moreover, until now, the wireless industry had mostly focused on raw bandwidth. However, the 5G networking debate was eventually moving beyond the tired discussion of raw speed and was starting to focus on pervasive connectivity to lay grounds for a fast and resilient link to the Internet whether a mobile user was in a subway train, at the top of a skyscraper, or in an exhibition center.

According to a 2011 study from Morgan Stanley, the growth curve for the mobile Internet could be around twelve times as steep as for the desktop Internet, which we remember how transformative that was during the 1990s. The report called the speed of the mobile Internet take-up a revolution the likes of which people haven't seen before. Mobile phones were reaching into the furthest corners of the world, and according to an ITU study, at the end of 2011, there were as many mobile phones as people inhabiting the planet: nearly 6 billion. Though most of these handsets were feature phones with limited capabilities, during this decade, the technologies would become so cheap that virtually every phone sold would become what the industry called a smartphone in 2011.

And every one of these phones would be constantly connected to the Internet. So what would happen when most of the inhabitants of this planet carried a gadget that gave them instant access to pretty much all of the world's information? The implications for almost every aspect of life were dizzying. Just imagine a world where everyone using the web on their mobile phones, and how this combination of Internet access and mobile personal computing devices could transform education, economic development, commerce, government services, and more. The good news was that the basic vision of the mobile Internet was in place by the mid-2010s. With a steady rise in network speed, it was hoped that the dream of pervasive wireless network would turn into a reality and that smartphones would make mobile data services as easy to use as the ubiquitous telephone system.

The web was going mobile not just in wealthy countries but across the world. The advent of inexpensive smartphones and tablets, linked to intelligent cloud-based services, was changing everything—including entertainment, medicine, education, and social development—in ways that would benefit all. The price of an Internet-capable phone had fallen below US$50 by 2014, and in India, it was possible to get tablets like the Aakash 2 for half of that. It was also cementing the dominance of web behemoths like Facebook, who had created stripped-down versions of its site, which could be used on a basic feature phone. The companies like Facebook were also persuading mobile operators to give users access to their mobile sites for free.

In mobile Internet's brief but exciting history, the defining moment came in 2007 when Apple took its largely unfinished personal digital assistant (PDA) business that it had started with Newton MessagePad and effectively turned it into a smartphone that could make calls and most importantly offered excellent web-browsing ability. The form factor of the iPhone did strike some resemblance with the all-screen look of the Newton. Fourteen years after Apple had released the Newton—a handheld device with handwriting recognition, desktop syncing, and an embrace of third-party applications—the Cupertino, California–based company was proudly showing its descendant hailed by the industry as a revolutionary and magical product.

Before the iPhone, a user could access the Internet after sorting through a dozen clicks to see mobile-optimized feeds on a tiny screen. For the wireless industry, data was only useful to the mobile enterprise workers and early adopters who used smartphones with clunky browsers that made mobile web surfing far less appealing. The iPhone refused to rely on the "baby Internet," as Jobs put it during his Macworld 2007 keynote address, and instead featured a mobile version of its own Safari web browser with tap- and pinch-to-zoom for an elegant, unprecedented browsing experience along with a powerful e-mail feature.

Apple not only launched the most compelling mobile experience for access to the web, it also defined the smartphone battle by creating the biggest and most active ecosystem of mobile apps. It rallied software developers behind a solid operating system based on a subset of its well-known desktop software. Apple's seductive phone with its powerful software and intuitive interface encouraged users to go online and stay online. The iPhone users consumed an average of up to ten times the bandwidth of mobile subscribers. They were playing games on their phones, sending video messages, and downloading music. A new network regime was taking shape in a wireless order created by smartphones like the iPhone.

The iPhone and Android pushed the mobile Internet well into the mainstream with hundreds of millions of subscribers. The wireless industry—then focused on voice and messaging services—was caught unguarded by the explosive growth in mobile Internet traffic. Nevertheless, the old and new mobile phone establishments were now betting that enough of the things that people wanted from the Internet could work without a personal computer's big screen. Despite the smaller display of cellular handsets, users were still able to send and receive e-mails, perform e-commerce transactions, and get real-time updates on such things as travel, weather, news and sporting events. Now tech pundits claimed that, with wireless Internet, computing would finally achieve its apotheosis: simple, reliable, ubiquitous, and pervasive.

1 THE NEXT WEB

"In ten years, there will be 50 billion devices connected to the web."
— Hans Vestberg, Ericsson's president and CEO, in an interview in April 2010

In 2013, Verizon installed as many as seventeen new radio base stations to extend its Long Term Evolution (LTE) wireless coverage across Montana and North Dakota, one of the fastest growing oil regions in the United States. However, Verizon wasn't expanding its 4G wireless network for potential smartphone users in this rangeland. Instead, the largest U.S. wireless operator was there for machines, positioning its mobile infrastructure as a network for drilling companies' Internet-enabled smart sensors and pumps. In the drilling areas, many work sites were off the grid and beyond the reach of ordinary phone lines; so the leading U.S. mobile operators were providing a viable option: mobile connections.

Mobile phone operators like Verizon were modernizing their wireless networks through the brand new LTE wireless technology amid the prospect of billions of devices joining the mobile Internet bandwagon in the coming years. Verizon, who had just recently built its LTE-based turbo-speed network, had also added low-cost sensors and mini radio transmitters to its network recipe and made its wireless infrastructure more attractive for new enterprise customers like oil companies. Verizon's foray into the oil fields of Montana and North Dakota was part of its efforts to reinvigorate revenue streams that were being increasingly commoditized in the cut-throat consumer wireless markets.

There were other important use cases. Take Holland, Michigan–based technology solutions provider Twisthink who used Verizon's wireless network to allow forklift operators remotely advance a rider pallet lift truck through a glove with a built-in radio transmitter. That way, Twisthink helped material-handling equipment companies reduce a forklift operator's footsteps by 70 percent a day, and saved labor time and cost. Then, there were agriculture forms using LTE network-hooked sensors to protect crops through the monitoring of water, air and soil data, and do a host of other things like detecting pregnancy in livestock or monitoring milking frequency.

One of the best kept secrets in the wireless industry during the early 2010s was that mobile operator's machine-to-machine communications (M2M) business in developed markets like the United States had been growing about 20 percent per year amid declines in the price of the M2M devices and data plans. Pundits called the Internet of Things—otherwise known as M2M communications—the biggest opportunity in the history of technology business. General Electric (GE) had coined the term Industrial Internet for this phenomenon while IBM had thousands of programs underway for the same idea under its Smarter Planet initiative.

In fact, among all these manifestations of the ubiquitous Internet, M2M was a more commonly used maxim in the communications world. However, this emerging network of the twenty first century comprised of links between human to human, human to machine, and machine to machine. The Internet of Things was the manifestation of the development and adoption of the consumer Internet in which an open, global network connected people, data, and machines. It was a natural outcome of the increasing computerization of all the devices. The Internet and humanity had converged back in the 1990s, and now it was the turn of the devices around humans to converge with this remarkable technology platform. Apparently, the mid-2010s were poised to be a breakout moment for this avant-garde phenomenon.

It wasn't just the availability of turbo-speed LTE networks and the ubiquity of wireless and broadband Internet access that had brought the idea of ubiquitous Internet near to a tipping point. The economy of scale of sensors, semiconductor chips and software programs that tracked movement and information flows was also a vital part of the next web that was transforming the entire IT landscape. The staggering growth of cloud computing and the advent of a new industry dedicated to cutting-edge data analytics were the other prominent catalysts in this coming-of-age tale of pervasive Internet. Finally, and probably most crucially, the rise of mobility via smartphones had overcome a plethora of barriers in the way of M2M through application programming interfaces (APIs) and open software platforms.

The notion of the Internet of Things began to gain wider industry recognition when the automated communication of data between connected devices started moving beyond use in utilities, transport and heavy industry, and into the mainstream consumer and communication environments. Here, before going any further into the Internet of Things labyrinth, it would be worthwhile to take a peek into the brief history and evolution of M2M or the Internet of Things.

RISE OF THE MACHINES

The Internet of Things was not a new idea. In 1988, Mark Weiser, a technologist at the Computer Science Lab of Xerox Palo Alto Research Center (PARC), put forward the notion of ubiquitous computing as information technology's next wave after mainframe and personal computers. In this new world, what he called "calm technology" would reside around us, interacting with users in natural ways to anticipate their needs. Weiser coined the term "ubiquitous computing" to describe a future in which personal computers would be replaced with invisible computers embedded in everyday objects. He believed that this would lead to an era of computing in which technology, rather than panicking people, would help them focus on what was really important.

In his article "The Computer for the 21st Century" published in *Scientific American* in 1991, Weiser wrote, "The most profound technologies are those that disappear. They weave themselves into the fabric of everyday life until they are indistinguishable from it." He continued his explorations into the idea of ubiquitous computing till 1998, a year before losing his battle to cancer. Weiser had set up a water fountain outside his office whose flow and height mimicked the volume and price trends of the stock market. "Ubiquitous computing is roughly the opposite of virtual reality," Weiser wrote. "Where virtual reality puts people inside a computer-generated world, ubiquitous computing forces the computer to live out here in the world with people."

Weiser's work—based on research on human-computer interaction and PARC's earlier work on computing—initially sparked efforts in areas such as mobile tablets and software agents. Subsequently, these efforts morphed into pursuing intelligent buildings packed with wireless sensor networks and displays, where information follows wherever people go. Weiser's vision was shared by many in the PC industry. The first practical manifestation of ubiquitous computing emerged in the early 1990s when John Doerr, the legendary venture capitalist at Kleiner Perkins Caufield & Byer, started the pen-computing frenzy by funding Go Corp. By 1991, the pen-based computing wave had become the "next big thing" in technology world. Yet, despite this pen-based computing rush, only a single product became commercially available from GRiD Systems, a small computer outfit on the east of the San Francisco Bay.

But then Apple Computer's chief executive officer, John Scully, fanned the flames of pen-based computing in a speech about a handheld computer he called the personal digital assistant or PDA. "Palmtop computing devices will be as ubiquitous as calculators by the end of this decade," he told his audience. Scully echoed Weiser's vision touting that computing would eventually go a step farther in the journey that started from mainframe to minicomputer to personal computer. In May 1992, Apple CEO announced the Newton, an amazingly ambitious handheld computer. Scully set the computer world on fire with his prediction that PDAs such as Apple's Newton would soon contribute a trillion-dollar market. He professed that this gadget would launch the "mother of all markets."

Apple's overwhelming electronic gizmo was built around Scully's vision of handheld computers. Representing a new class of portable devices, the Newton was stitched together from several emerging technologies to perform specific tasks like memos and personal diary. It was capable of sending and receiving messages, and was equipped with modest computing and handwriting recognition capabilities. In August 1993, when Apple introduced the Newton, Scully foresaw a world in which a simple device would connect mobile users to vast networks of information. However, Newton was not only ahead of the market; it was also fatally flawed from technological and commercial standpoints.

It had the feel of an electronic brick both in size and in weight. A user could barely hold it in hand. Newton failed to connect with the rest of the computing world, and five years after its launch, the newly arrived chief executive Steve Jobs abandoned the product to focus on Apple's core Macintosh lineup. But the Newton debacle proved a kind of start-over that led to a new generation of PDAs that would focus on more practical features. A plethora of such products sprang up—offering some sort of interactive capability—and among them was Palm Pilot, a simple, no-frills compact device which hit the market in February 1996. Palm Pilot became one of the fastest-selling high-tech toys of the decade. The elegant little computer became an American icon; one million Palm Pilots were sold in the first eighteen months.

PDAs continued their slow and modest journey toward gaining interactivity and getting assimilated into the network. Meanwhile, a British technologist Kevin Ashton became interested in incorporating radio frequency identification (RFID) chips into products with smaller form factors while he was working as an assistant brand manager at Procter & Gamble in 1997. After a couple of years' of work, he proposed using RFID chips to help manage P&G's supply chain problems. He argued in an article that having humans input data was incredibly clumsy and inefficient. On the other hand, semiconductor chips and sensors were becoming smaller, cheaper and less power hungry, so they could be incorporated into just about anything.

Ashton suggested that getting information from objects themselves could revolutionize the supply chain. The "Internet of Things" was born. Ashton recalled the episode for the *RFID Journal* later in 2009, "I could be wrong, but I'm fairly sure the phrase "Internet of Things" started life as the title of a presentation I made at P&G in 1999. Linking the new idea of RFID in P&G's supply chain to the then-red-hot topic of the Internet was more than just a good way to get executive attention. It summed up an important insight which is still often misunderstood." The use of an RFID chip within a miniature device connected wirelessly was akin to a simple SIM card, and it expanded the reach of the Internet of Things to healthcare, automobile, energy, and more.

Ashton's work eventually led him to MIT, where he helped start an RFID research consortium called the Auto-ID Center with professors Sanjay Sarma and Sunny Siu and researcher David Brock. The center opened in 1999 as an industry sponsored research project with the goal of creating a global open standard system to put RFID devices everywhere. Ashton was the Center's executive director. Another industry luminary who popularized the underlying concepts related to the Internet of Things was Neil Gershenfeld of the MIT Media Lab. He wrote in his book *"When Things Start to Think"* published in 1999 that, "In retrospect it looks like the rapid growth of the World Wide Web may have been just the trigger charge that is now setting off the real explosion, as things start to use the Net."

Next year, in 2000, LG Electronics announced plans to launch the first Internet refrigerator. Later, in 2005, the notion of the Internet of Things got official recognition from the communications world when the International Telecommunications Union (ITU) published its first report on this emerging industry discipline. The report acknowledged, "A new dimension has been added to the world of information and communication technologies (ICTs): from anytime, anyplace connectivity for anyone, we will now have connectivity for anything. Connections will multiply and create an entirely new dynamic network of networks—an Internet of Things."

However, according to Cisco's Internet Business Solutions Group (IBSG), the Internet of Things was born between 2008 and 2009 simply at a point in time when more "things or objects" were connected to the Internet than people. The study claimed that the number of devices connected to the Internet reached 12.5 billion in 2010 amid the exponential growth in the number of portable devices like smartphones and tablets, while the world's human population came down to 6.8 billion, making the number of connected devices per person more than one for the first time in history.

There were millions of machines across the world, ranging from power grids that produced electricity to airplanes that moved people and cargo around the world. In the healthcare segment, for instance, tiny devices worn by patients provided real-time medical data and enabled the dispensing of medication. In utilities, M2M communications provided near-real-time data to consumers on their usage through smart meters. Next, in the transport and logistics industry, pallets and packages were able to communicate their location, allowing for real-time parcel tracking. There were thousands of complex networks ranging from power grids to railroad systems that tied machines and fleets together.

Clearly, this vast physical world of machines, facilities, fleets and networks could more deeply merge with the connectivity, big data and analytics of the digital world to form the Internet of Things. Among these network-centric innovations, the most prominent early contributions came from mobile phone carriers who had been using their 2G and 3G networks to connect everything from jukeboxes to ice machines since the late 1990s. The use of the mobile technology as a payment gateway had started in Helsinki in 1997 when a company owned by Coca-Cola installed two mobile-optimized vending machines. These machines accepted payment via text messages.

The cellular-centric M2M communications industry had emerged in 1995 when Siemens set up a dedicated department inside its mobile phones business unit to develop and launch a GSM data module called M1. It was based on the Siemens mobile phone S6 for M2M industrial applications and enabled machines to communicate over wireless networks. In October 2000, the modules department formed a separate business unit inside Siemens called "Wireless Modules," which in June 2008 became a standalone company called Cinterion Wireless Modules. The companies like General Motors and Hughes Electronics were also among the early implementers of the M2M technology.

Over the course of a decade, while the cost of building a wireless network drastically came down, the wireless coverage, speed, and capacity increased significantly. So, mobile phone operators, who had long dreaded of becoming dumb pipes for apps and service providers like Apple and Google, started contemplating a second life for their networks as the backbone of the Internet of Things. All of those connected cars, meters, healthcare gizmos and wearable gadgets would need an always-on ubiquitous network, and mobile phone networks seemed ideal for the task.

Moreover, by the year 2014, when mobile phone penetration had neared 100 percent, the growth of companies like Verizon seemed dependent on innovative new M2M services. It had become apparent by now that at the center of the Web of Things was the smartphone which people carried around during most of their waking hours. The smartphone of 2014 boasted more processing power than the Apollo program which put a man on the moon and had become a universal remote control of the digital life. Case in point: LG Electronics was launching smart appliances using the Android smartphone platform as the operating system and in-home Wi-Fi as the connection to the Internet.

The notions of connected home and the Internet of Things had remained on the fringes for a decade. So what was it that finally catalyzed M2M communication markets and made the Internet of things a commercial reality in 2014? In the final analysis, the Internet of Things, after a decade of talking, finally became a driving force with the advent of devices like the iPhone. In fact, it was the iPhone which became a catalyst for the growth of the Internet of Things beyond the luxury and industrial markets. It was the iPhone, for instance, that Pacific Gas and Electric Company (PG&E) was employing to control thermostats through smart grid apps. The origin of the phone becoming a sensor platform and eventually the central nervous system could also be traced in the making of the iPhone.

THE FOURTH IT WAVE

Nineteen years after Mark Weiser envisioned the concept of ubiquitous computing and fourteen years after Apple's ill-fated attempt to turn this dream into a reality in the form of the Newton, its descendant, the iPhone completed the unfinished business, translating the vision of ubiquitous computing into a commercial proposition. In its very essence, ubiquitous computing needed a ubiquitous communications system, and the iPhone just filled that vacuum in 2007. The mobile Internet reached an inflection point from where it would start its quest to become ubiquitous. It was the notion of Internet ubiquity that promised a world based on the certainty that wherever someone was, he or she was constantly connected to the Internet and its services. And once Internet access became truly ubiquitous—as taken for granted as electric lights or running water—it would become as much a fundamental part of the infrastructure as bridges, roads, and tunnels.

Nokia put out some of the first GPS phones and then came Apple's iPhone, a masterpiece with gesture-based multitouch interface, an elegant web browser and sophisticated music and video playback capability. The unique combination of the mobile Internet connectivity, GPS, and the ubiquity of handheld devices would open the floodgates of new, cost-effective ways to access information. To Apple's credit, it reinvigorated mobile handsets through innovative user interface, touchscreen and powerful add-on apps, and to Google's credit, it provided a similarly powerful solution on a semi-open platform. Otherwise, the first two generations of iPhone competitors were blatant failures because of the poor usability created by their uncoordinated user experience. It's not enough to have a nice touchscreen or an app store. The integration of hardware, user interface, and apps functionality had to come together to form a supportive whole. Individual apps could succeed only by nicely fitting into a platform.

The iPhone was the first truly functional touchscreen phone. Apple realized that the mobile phone was becoming the default gateway for people to experience content: games, movies and the web itself. But existing phones with their small screens, cramped keypads, and legacy operating systems were doing a poor job on this count. The iPhone solved these problems by combining a large, high-resolution glass touchscreen with a button-less user interface. Moreover, Apple introduced the element of emotion into the user interface by mimicking the way humans work in the real world—sliding a switch, flicking through pages, allowing objects to be held or manipulated in a tangible manner. By using real-world metaphors, touch interfaces brought the physicality of interaction to the fore and made technology simpler and easier to relate to for end-users. Once touchscreen devices did away with constraints archetypical to contemporary gadgets, even kids and older people could comfortably use mobile phones.

Apple harnessed another revolution by equipping the iPhone with an accelerometer to switch its display automatically from portrait to landscape orientation. The accelerometer's ability to respond to a user's motions turned previously pedestrian operations into game-like experiences. Apple was able to create novel apps by exploiting the accelerometer for gaming, health monitoring, sports training and countless other uses thought up by legions of third-party software developers. By using sensor fusion, these software developers could take information from all these sensors and create apps that had never been thought of. Apple even dedicated the M7 chip in the iPhone 5S handset for processing motion data from the phone's accelerometers, gyroscope, and compass sensors.

The addition of sensors to the computing-mobile phone combo enabled new platforms and enhanced the magic of connectivity. Smartphones got smarter because of all the sensors being added to them; having a mere Internet connection didn't make a phone smart. Sensors—such as gyroscopes, barometers, accelerometers and others—were what enabled connected devices to collect useful data about our bodies and the environment. Sensors enabled smartphones to recognize gestures and provide driving directions. They enabled new applications with context awareness that promised to offer personalized, relevant content by sensing the environment, monitoring user-states and adapting a more intuitive experience. The truth was that the ever-expanding ecosystem of smartphone apps owed a great deal to the micro electro-mechanical system (MEMS) sensors.

Take healthcare, for example, where increasing adoption of electronic health records was making it easier to incorporate digital information from smartphones into doctor's visits. Apps available on the iPhone would let doctors remotely diagnose patients, tapping cell phone sensors as diagnostic devices. With just an hour's notice, a service called 3GDoctor let patients consult with a doctor via a 3G videophone connection. An automated assistant collected the medical history and information on the symptoms, including, for example, an image of a wound or an audio sample of a cough. A doctor remotely evaluated the information, texted the patient to ask questions or request further images or audio, and then called the patient to discuss the diagnosis and make recommendations for treatment using the face-to-face capability of videophones that had both front- and rear-facing cameras.

Smartphone was inherently a sensor now. So, just like the apps envy, smartphone makers were scrambling to match Apple's sensor complement. The iPhone had opened the eyes of handset vendors to how a sensor—like accelerometer—could harness motion. The motion sensor in the iPhone provided users a simple way to read wider web pages simply by allowing them to turn the handset sideways to rotate the display. Now mobile phone makers didn't have to identify the next killer application, they just needed to add the sensors—the invisible computers—and apps would find the best way to use them. For example, combining motion sensors with location-based technologies and payment apps could create a new set of opportunities for people to interact with phones.

A number of smartphone units now came equipped with sensors to record movements, sense proximity to other people with phones, and detect light levels. Smartphones, with their always-on Internet access, were on the way to becoming the world's premier wireless sensor network. Eventually, the race to add accelerometers to the early generation of smartphones turned into a pursuit for incorporating MEMS-based gyroscopes, and next in line would be the addition of barometric pressure, humidity, and temperature sensors. Accelerometers gave users only a crude measure of motion, but with the addition of gyros and other sensors, they could perform true motion processing. Smartphone-based wireless sensor network capable of monitoring air quality, for instance, could become a higher-resolution network for toxic gas detection in a packed stadium.

Apparently, wireless industry had only scratched the surface of applications to which sensors could be put to use. Eventually, thousands of smartphone apps took advantage of user's location data to forecast traffic congestion and rate restaurants. The next generation of apps would take advantage of unique sensors on devices and networks intertwined with 3D cameras, multiple microphones, and increasingly precise GPS to take smartphones to a new level of user experience. Adding this extra layer of sensing on top of the mobile Internet could eventually facilitate the creation of the fourth IT wave. The IT industry has passed through the initial three information technology waves—mainframe and mini computers, personal computers, and portable networked computers—and now it was on the verge of the fourth wave, that of "IT everywhere."

MOBILE INTERNET OF THINGS

In the 1990s, the concept of remotely monitoring and controlling distributed assets and devices was mostly reserved for large and expensive investments like power plants and dams. Fast forward to 2013, connected products were expanding to e-books, cars, home appliances, smart grids, manufacturing, fast food, security, healthcare, and more. The higher capacity LTE networks offered the promise of transforming vehicles and other entities into mobile communications centers, providing heightened security, infotainment and a host of outbound data flow. By 2020, billions of things—from clothes to cars and from body sensors to tracking tags—were forecast to be connected to mobile networks. That could consume 1,000 times as much data as mobile gadgets of early 2010s, so while mobile operators rushed to roll out 4G networks, the wireless establishment was already beginning to define 5G wireless standards.

Mobile carriers continued to tinker with consumer data pricing in the post-iPhone era, but they had also started to think beyond smartphones and focus on the Internet of Things, connecting everything from tablets to cars to home appliances. AT&T and General Motors, for instance, planned to embed LTE technology into millions of cars as an upgrade to GM's OnStar infotainment service. GM provided the in-vehicle OnStar service to nearly 5.5 million car owners in North America who were willing to pay for this subscription-based service and supply data on their whereabouts in exchange for safety and convenience. AT&T was also working closely with Audi and Tesla Motors to develop web-connected car technologies.

Earlier in 2006, the second largest U.S. mobile carrier had taken full control over its wireless business, and now it aspired to facilitate digital life for its customers through ideas like connected home and connected car. In 2009, AT&T announced a strategic alliance with Jasper Wireless to jointly support the creation of M2M devices. Next year, the Ma Bell descendant joined hands with KPN, Rogers, Telcel and Jasper Wireless to collaborate on the creation of an M2M site, which would serve as a hub for app developers. Likewise, Sprint had entered into a strategic alliance with Axeda Corp. for developing M2M solutions.

Also in 2010, Verizon had consolidated its strategic alliance with nPhase—a joint partnership of Qualcomm and Verizon—to offer enterprise customers an easy way to roll out M2M solutions across Europe and the United States. AT&T's archrival Verizon was also among the companies who believed in this brave new world of connected cars, kitchen appliances and wearable gizmos, and was trying to lay grounds for this new business through a greater synergy between its wired and wireless networks. The largest U.S. wireless carrier had pioneered the smart home remote monitoring service in early 2012. The smart home service cost US$10 a month, in addition to a starter hardware kit, which cost US$130, and provided a gateway device to connect to radio modules that in turn were connected to surveillance cameras, appliances, lights or thermostats.

The radio modules embedded in objects like surveillance cameras and appliances were about the size of two thumbnails and consumed so little power that they could remain connected for as a long as twenty years before their module batteries would require recharging. Verizon required either a fiber optic link or copper wired broadband connection to the home for the service, which in turn, worked over Z-Wave wireless technology inside the home. The gateway used in the service combined Z-Wave wireless links with broadband Internet connection to allow users to monitor and control home systems from smartphones and other devices.

Mobile phone operators had long been fearful of becoming the dumb pipes of raw bandwidth. The Internet of mobile things was their opportunity to seize the moment and manage connections and services in the new wireless order. In 2014, mobile operators were demonstrating connected home and connected city applications, sticking 4G wireless technology-based LTE radios in smart utility meters and public transit and healthcare devices. In fact, the whole idea behind the futuristic 5G was to build ubiquitous networks offering plentiful and cheap mobile data and create a more efficient life for people.

The fundamental concept of 5G wireless demanded the network to be not a monolithic entity. One of the key merits of the 5G networks was going to be their ability to handle billions of connected devices and myriad of traffic types. In other words, what was transformative about the 5G technology was its ability to offer different capabilities for different traffic types: it could serve the industrial Internet and Facebook apps at the same time. That's why the key elements of these futuristic networks spanned from smart antennas to ultra-dense deployments and from improved coordination between base stations to device-to-device communication.

Mark Weiser's work on "ubiquitous computing" at Xerox PARC back in 1988 had conceived wireless data as the underlying technology for a computing environment that enabled personal mobility of computer users. Now some people in the wireless industry believed that the fourth generation of wireless networks—4G—embodied that vision of "ubiquitous computing" by offering mobile users the ability to access the applications from any platform, anywhere, any time. However, to create such an environment, the industry needed to integrate various applications working in harmony with intelligent sensor networks. For instance, users' car sent SMS to their mobile phones if someone tried to open the door while they were away from their cars. Or a user's home security camera was hooked to the Internet so that he or she could view the living room on his or her mobile phone screen.

Things were changing in the mobile realm. The smartphone had marked a new era in portable electronics with connectivity and that connectivity came via the Internet. With the power of the mobile Internet on its side, smartphones were now establishing themselves as personal gateway to the wealth of information available on the web. The coming together of mobile phones and the Internet was going to unleash the next tidal wave in consumer electronics that pundits had been talking about for almost two decades. Truly ubiquitous wireless broadband would be much more than a mere combination of media and communication platforms. It could become the nervous system of civilization, enabling a thin mesh of always-on smart devices to coordinate and facilitate almost every aspect of human life.

The mid-2010s had also marked the beginning of the era of connected wearables in which the industry was steadily working to overcome limitations in battery life, network connectivity and form factor. The wearable industry was kicking off with some of the compelling use cases: clip-on heart monitors, glasses, smartwatches, wristbands, etc. Wearable computers were also being embedded into objects like cars and homes. And all of this was happening at the cross-section of mobility, cloud computing and big data. Google Glass was a classical example of such an amalgam of new technologies that promised to define the next generation of pervasive computing.

Moreover, beyond this maze of billions of connected devices with potential to join the mobile Internet juggernaut, there were also billions of humans waiting to join the smartphone revolution. A US$100 smartphone was amazing in the sense that it would allow billions of people to connect to Facebook, Google, Twitter, medical resources, and learning and job opportunities on an actual personal computer at a price once thought all but impossible. Smartphone—with the power of wireless web on its side—was no more a status symbol; it was a productive and creative communication tool used by more than a billion people seeking to improve their work, increase their wealth, add to their joys and connect to the rest of the world.

Meanwhile, software developers continued to build mobile websites that could work across any screen size. The so-called responsive design was likely to continue to grow beyond 2014 because the old websites were losing businesses a lot of traffic from mobile devices. Last but not least, the arrival for super-fast LTE wireless networks was about to turbo-charge an already exhilarating wireless web scene. In mid-2010s, after more than a decade of ups and downs, the crude version of the late 1990s mobile Internet had come a long way. The mobile Internet had become the business of the future at a time when not many people were even talking about it.

People were talking about the Internet of Things and its different manifestations. However, a closer look revealed that the Internet of Things was the next incarnation of mobile Internet, a forerunner of sorts. Here, at this point, before we go any further and explore this new embodiment of digital life, it would be worthwhile to scroll back and see how this great technology undertaking kick-started in the first place and evolved over the course of a decade. The history and evolution of the wired Internet have been well documented, but the genesis of mobile Internet isn't that well known. The next couple of chapters attempt to provide an authentic account of wireless Internet's humble beginnings.

The notion of ubiquitous computing that Mark Weiser (sitting on left) conceived in the late 1980s eventually touched the commercial periphery nearly two decades later in the form of wearable computing devices like Google Glass, Nest thermostats and Proteus heart sensors. The image is taken from his 1991 article "The Computer for the 21st Century."
Image: *Scientific American*

Kevin Ashton, showing an RFID chip in this photo, is known to have coined the term Internet of Things back in 1999. The British technologist later told *RFID Journal* that he used the phrase "Internet of Things" to get P&G managers' attention during a presentation about RFID-centric supply chains.

2 PRE-iPHONE MOBILE WEB

"If you understand the concept of i-mode, you can succeed anywhere."
— Keiichi Enoki, managing director of NTT DoCoMo's iconic mobile Internet service

Nokia's Anssi Vanjoki first heard about the Internet in 1993 at a time when his company was heavily betting on digital cellular phones. One day Vanjoki saw a new hire hunched over a strange-looking database on his personal computer. It turned out the newcomer was online, using the Gopher menu system to browse through a library at the University of Texas. Vanjoki thought that if he could do this on the PC, he should be able to do the same on a digital handset. Across Scandinavia, many engineers were having similar thoughts. It might have seemed implausible at that time that people would surf the Internet and exchange e-mail messages from their cell phones given their tiny screens and awkward keypads. About six years later, mobile phone users were unmistakably doing so and in reasonably large numbers.

The notion of the mobile Internet first emerged on the technology scene during the mid-1990s when the mobile phone was on its way to ubiquity. When dreamers started talking about accessing the Internet from cellular handsets, the mere idea of coalescing two of the most transformational technologies of the twentieth century sounded fascinating. Science fiction writers had long been dreaming for a communication upheaval as a means to perfecting the society through wearable devices like the *Star Trek* communicators and the *Dick Tracy* watch phones and that notion was actually accelerating toward reality by the end of the twentieth century. Cellular phones and PDAs—already smaller than the *Star Trek* communicators and Captain's electronic logs—were starting to bring the world on the verge of a highly mobile society.

The wireless euphoria of the 1990s ended with a sense of enchantment. The mobile phone's triumph over the fixed-line telephony as the primary means of voice communications seemed assured by now. It was the sudden arrival of data, however, which was rewriting the rules of the mobile communications game. After cellular handsets became nearly as commonplace as a wallet, the telecommunications industry found itself at a new crossroads where a unique combination of mobility and connectivity promised a whole new world of possibilities. In fact, the natural progression of wireless data technology was only a few years behind its wired-line counterpart: the Internet.

Although an exhilarating phase of growth was just around the corner, the fabric of the mobile data market was far more complex than the traditional cellular settings that made up the voice-centric business model. Such an ordeal, however, applied to almost every technological endeavor, so hardly anyone was anticipating that the realization of data communications in the wireless space would happen overnight. Historically, mobile communications had focused on handling simple telephony services. The first two generations of cellular systems were designed purely for voice communications.

Wireless data applications had been in place for quite some time in specific industry areas and vertical markets. Vehicle dispatch services were a prime example. However, for consumers and high-end users, a new plateau was required to close the gap between the wireless and the data networking worlds. Short message service (SMS), through its mass-market appeal, had provided the initial impetus to wireless data delivery. Now the industry was gradually coming to terms with a new paradigm that would transform wireless data into a new market dynamo: they called it the "mobile Internet." And reminiscent of SMS's role in popularizing wireless data transport, the crystal ball for the mobile Internet came from a place the industry gurus were looking the least.

However, before we look into that, here is a little recap of the history. The evolution of the mobile Internet could be classified into three classical episodes, and they spanned across three continents acquiring distinct spaces as well as time frames. The fact that Apple and Google subsequently became the key figures in the mobile Internet coming-of-age story speaks well about tech-happy America's bragging rights on smartphones. After all, Americans not only invented the Internet but also much of the mobile telephony. In retrospect, the U.S. companies figured out how to seize the moment in mobile Internet trajectory, partly because smartphones incarnated from computers and partly because the mobile Internet was the inflection point of two great technology stories that America helped build in the first place: the Internet and mobile telephony.

However, in the heydays of digital wireless boom during the 1990s, Europe had gotten ahead of America in mobile telephony precisely because of its successful policies. While the United States left its cellular systems to the forces of the market and wound up with a fractured system, Europe settled on a single technology standard, Global System for Mobile Communications (GSM); charged little for licenses; and raced ahead. When Europe became the capital of the wireless universe in the 1990s, the notion of the mobile Internet started to be seen as Europe's best chance to extend its wireless lead into cyberspace.

The Europeans thought that while they had lost out on the first round of the Internet, they could catch up and even leapfrog

America with the mobile web. In the 1980s, Europe missed on the PC revolution as Americans creamed the European protected national champions such as Bull, Nixdorf, and Olivetti. In the next decade, the continent lagged behind in the Internet buildup. So for Europe, the mobile Internet was yet another rich opportunity: it offered the continent its best chance to launch its own tech-driven New Economy. Mobile Internet, Europeans said, would create a new wireless world and perhaps even give birth to another economic miracle after GSM.

However, at the dawn of the new century, the action was gradually shifting to Asia, which emerged as a new ground zero in the wireless wars. Much of the credit went to NTT DoCoMo who beat its European rivals out of the starting gate with i-mode. The next wireless course now seemed to belong to Asia where countries such as China, Japan, the Philippines and South Korea were increasingly becoming mobile phone zealots. The firing shot on mobile Internet also came from the Land of the Rising Sun when DoCoMo created an island of success in the form of i-mode service by adopting the wired Internet-like business model.

THE JAPANESE TAP DANCE

In early 1997, NTT DoCoMo president Koyji Ohboshi called the managing director of business marketing, Keiichi Enoki, in his office and asked him to launch Internet services over mobile phones. Personal computers in Japan were expensive, and homes were small, which left the Japanese far behind the United States in the PC arena. Japan was slow in its uptake of personal computers for several reasons, but one large factor was that the early desktop machines were difficult to use for those not fluent in English. Moreover, people had to commute in trains for a number of hours, so they had relatively less time to use the desktop at home as a portal to the Internet. DoCoMo saw a window of opportunity here. The company set a new mission: to lead Japan with the world's first instantly accessible mobile Internet service, i-mode.

Enoki instantly set up a new department that would aim to develop a range of free and pay-per-use services: send 500-byte e-mails, do banking, reserve tickets, and buy and sell goods. Mobile phone users could also check restaurant guides, phone directories, dictionaries, weather, stock prices, news, horoscopes, music charts, as well play games and download cartoon characters to their handsets. In its carefully crafted marketing campaign, DoCoMo emphasized convenience and avoided phrases like "advanced functions" and even the word "Internet." Mr. i-mode, as colleagues would call Enoki, also decided not to alter the shape of the handset, realizing that mobile phone users in Japan were used to having small and cute gadgets. The i-mode service would be fashionable yet functional.

DoCoMo won a million i-mode customers in less than six months. The Japanese people could trade stocks, make reservations, and check headline news by paying for the data they transmitted or downloaded through i-mode phones. Consumers loved the simple-to-use service and quickly adopted it for displaying Hello Kitty on their color screens and for diversions such as online gaming. For business people, i-mode brought the convenience of being able to use the phone like an office. Especially for people who didn't want to sit at their desks, it was a blessing. For others, i-mode simply meant fun. Teenagers, for instance, found their own uses of i-mode. DoCoMo had collaborated with multimedia behemoth Sony to bring games and cartoons onto mobile phones through i-mode handsets.

Within a few months after its launch on February 22, 1999, about 4,700 web pages had been designed for small handset screens and were made available to i-mode users. The screen-mode packet data service operating at a speed of 9.6 Kbit/s provided access to mini-websites through a specialized NTT DoCoMo portal. Each i-mode phone had a small button that granted easy access to thousands of DoCoMo-endorsed sites. Some were free; others charged up to US$2.50 a month. Now people stopped talking and started thumbing. By the turn of the century, thanks to i-mode magic, DoCoMo was winning a million new subscribers a month.

The i-mode service, through smartly tailored applications and limited web access, had established the world's first cradle of mobile web culture. In just two years after its launch, DoCoMo had attracted more than 22 million users and 30,000 content providers. The service became the epitome of what could be achieved in the mobile Internet space, the end-all, be-all of wireless data. DoCoMo's ordeal with i-mode was nothing short of inspiring; no wonder everyone in the industry wanted to know its secret. To understand i-mode's acceptance, it was necessary to look at the service from both the technological and the business-model perspectives. On the technology front, the Japanese wireless starlet decided to go for packet switching instead of conventional circuit switching technology being used in cellular networks. The i-mode technology was an always-on service: a user clicked a button and he or she was there. Packet switching saved users the hassles of calling up a website every time they needed to interact with one.

Then, the Japanese mobile operator came up with its own proprietary platform based on the Internet's standard script: HTML. That made it easier for developers to write applications without having to learn an entirely new computer language. The i-mode service used common HTML tags so that pages created for display on handsets could also be seen on the PC. Logica plc of London developed a compact form of HTML that it termed as cHTML; Access Co. Ltd, a Tokyo-based supplier of embedded browsers for non-PC products, developed the micro-browser for i-mode handsets. Access also conducted the overall technology integration in collaboration with Fujitsu, Matsushita, Mitsubishi, NEC, and Sony.

Nevertheless, the real genius behind the i-mode success was the way DoCoMo's business model optimized a relatively simple technology. What i-mode offered to end users was rich content; DoCoMo had established partnerships with hundreds of companies. The mobile phone operator offered an attractive pricing model to its partner websites. After collecting the fee and taking a 9 percent service charge, DoCoMo gave the remainder to the website. By giving content producers a means to charge users, DoCoMo ensured that there was plenty of content available. Takeshi Natsuno was credited for designing a profitable, info-entertainment platform that morphed into Japan's biggest market success in years. The i-mode multimedia guru lured high-quality content providers with a micro-payment scheme that enabled them to charge fees for their services—ranging from weather updates to horoscope.

JAPAN'S WIRELESS MIRACLE

Tokyo was now the capital of the mobile universe. The success of i-mode service had made DoCoMo the poster child of the wireless revolution. While other mobile phone operators were still trying to figure out the math, DoCoMo had shown the world how to make money from the wireless Internet. At the height of i-mode euphoria, DoCoMo buoyed the Japanese electronics industry—then swimming in the red—when it raked in more than US$18 billion on its initial listing of shares on the Tokyo Stock Exchange. In a matter of months, DoCoMo became Japan's largest capitalized firm with more than US$100 billion in stock value, which placed it ahead of Toyota and even its parent company, Nippon Telegraph and Telephone Corp. The pioneer of the mobile web also became the second most valuable wireless company in the world after Vodafone.

DoCoMo had won 32 million subscribers in just three years after i-mode launch. The i-mode service turned out to be a social phenomenon that transformed Japan forever. But the story of nation's ascent to wireless glory had a couple of intriguing twists. The country had been in a habit of protecting its markets from foreign influence through somewhat tricky regulatory arrangements. For instance, mobile phone users couldn't even own a handset; they had to rent it from the service provider. Eventually, under a strong pressure from the United States, Japan opened up its mobile phone market in 1994. As it turned out, that was a blessing in disguise.

Among the most notable achievements that Motorola's Christopher B. Galvin—who later became CEO of the company his grandfather founded in 1928—counted was having helped to open the Japanese telecommunications market to foreign mobile phone companies. Charlene Barshefsky, then the deputy U.S. Trade Representative, was reluctant to start a trade battle until the younger Galvin assured her that Motorola would share the heat if Japan resisted. As a result of their efforts, in the early 1990s, Japan permitted more foreign competition and began letting consumers own their cellular phones. Consequently, the country became the fastest-growing cellular market in the world, and by the end of the decade, the number of mobile phone users in Japan had surpassed 50 million. Motorola, however, could not make a breakthrough in selling handsets in Japan when Japan's love affair with mobile phones began.

Japan's manic sweep to the mobile Internet was also attributed to one of its great wireless failures: Personal Handyphone Service or PHS. Pushed by the government in the early 1990s, Japan's electronics and telecommunications industries tried to create a global standard for personal communications gadgetry. PHS—launched in Japan on July 1, 1995—functioned as a cordless extension at homes and offices and as a portable phone in the street for high-density pedestrian traffic. Initially a huge success in Japan, it boasted a data rate of 64 Kbit/s that was many times faster than standard cellular networks. But the success proved short-lived as PHS couldn't stand against the rising tide of mobile phone services. The technology that provided a cheaper alternative to cellular services flopped abroad and gradually shrank in Japan, but it provided the ground on which Japan could build robust new wireless platforms.

Also, earlier in 1991, as part of the Japanese government's cautious, or rather calculated drive to diversify the national telecom monopoly, NTT spun off its mobile phone unit. In the coming years, NTT Mobile Communications Network Inc. (NTT DoCoMo) turned itself into a futuristic technology company by adapting the speed and brutality needed to compete in the global marketplace. And by embracing new technologies, it quickly developed the future vision of a service-oriented company.

DoCoMo was the abbreviation of its corporate slogan "Do Communications over the Mobile Network." It also happened to be a Japanese colloquialism for the word *everywhere*.

The Japanese wireless operator embedded cell phones deeply into the local culture when it launched its phenomenally successful i-mode service. DoCoMo sorted out how to bring data services on handset screens when the mobile Internet was still a virgin territory. Its relentless focus and willingness to go to painstaking lengths to make life easier for consumers was what turned DoCoMo into an incredibly successful company. The world's most valuable success story in wireless web adoption was also credited to its senior executives and their un-Japanese willingness to challenge convention and to pursue aggressive strategies in meeting objectives. With i-mode, a service that had been viewed as an inflection point for the mass mobile Internet, the company was suddenly far ahead of its rivals.

Japan, a nation of compulsive dialers, had reached the pinnacle of the great wireless transition. Now DoCoMo was like a huge sumo wrestler overpowering the marketplace, and there was nowhere left for it to go but overseas. The company that made cell phones a must-have item in the now-frugal Japan aimed to become a formidable design arbiter of the next pervasive computing platform. So it started taking its expertise around the globe through relationships with mobile operators in America, Asia, and Europe. When its share price peaked in spring 2001, DoCoMo was the eleventh-largest company in the world with enough cash to buy a 16 percent chunk of AT&T Wireless, the third-largest mobile phone company in the United States.

THE WAP SAGA

The story of the mobile Internet wasn't all a win–win tale. Almost parallel to the making of i-mode, the wireless industry at large was at the helm of a much broader initiative to provide an early taste of the mobile Internet. At the time when Japan was thinking over how to bring cyberspace onto mobile phones, Ericsson and Nokia were also chalking out their own Internet strategy. The European cellular titans, however, were merely following a lead from a small American outfit.

The affair began in July 1996 when AT&T Wireless rolled out the PacketNet data service on mobile phones using a technology from Unwired Planet. The California upstart, led by an Apple Computer veteran Alain Rossman, had developed the microbrowser technology that enabled Internet access over mobile handsets. A year later, Unwired Planet, rather than go it alone, approached cellular industry heavyweights Ericsson and Nokia to persuade them that its mobile phone browser should become the standard for wireless Internet communications. Ericsson and Nokia agreed to join Unwired Planet on a forum for specifying WAP, short for Wireless Application Protocol. Motorola's last-minute joining to complete the wireless troika was reminiscent of its jumping onto the Symbian ship.

There were other parallels with Symbian. Just like Psion, Unwired Planet had gained a technological lead over large mobile phone firms and software rivals, so WAP was going to be an open standard rather than a proprietary one like Microsoft's Windows. It was quite unusual to see a little startup creating a de facto standard by turning its idea into a global initiative with support from major industry players. The company that was to give WAP to the world boasted to be the first ever wireless dotcom company when it changed its name to Phone.com at the height of the Internet bubble.

As for the handset business, by the late 1990s, it had become a monopoly of three wireless manufacturers: Ericsson, Motorola, and Nokia. Only Korean makers were able to break into the CDMA handsets market because, except for Motorola, these wireless magnates were not much focused in that particular segment. The question for the so-called wireless troika was how to share the market and maintain dominance in this flourishing but rapidly changing marketplace. A new industry segment was in the pipeline that renewed the hopes of the stagnant wireless data market. So the trio decided to acquire the role of the wireless industry's standard-bearer in a very subtle way.

After its inception in June 1997, WAP gathered a lot of support around the world as a technology-independent underlying link with transport delivery mechanism. It was designed for a wide range of applications on smaller screens of mobile phones with the focal point on economical uses of Internet resources and air-link capacity. The effective use of wireless bandwidth—a precious commodity in this case—was crucial in the realization of the mobile Internet. Another important goal was to rework the web into simple text that would be readable on small display spaces. For this purpose, wireless markup language (WML) was developed for wireless handsets; it was meant to be very similar to HTML.

The protocol was to cater wireless carrier-centric applications with WAP gateways performing encoding and script compiler functions for delivery of data onto handsets. The gateway servers were required to do the protocol conversion from HTTP and TCP/IP to WAP, and content conversion from WML text to binary data. The task was to be handled by mobile phone operators equipped with WAP servers; wireless manufacturers were only to produce WAP-enabled handsets. A small piece of software installed in a cell phone—a micro-browser—would help display data sent from a WAP server.

To stake full claim to the riches of the mobile Internet, the first WAP service was rolled out in the fall of 1999. By then the press was full of rosy predictions: WAP was going to take over the web, kill off the PC, and change everything. But Europe's phone companies flunked their first Internet exam. First, there were not enough handsets, and when the handsets finally arrived somewhere in 2000, users found themselves struggling with slow data speeds, primitive applications, and sky-high connection charges. Consumers didn't like it, wireless carriers didn't like it, and content companies didn't like it. The over-hyped WAP service failed miserably in Europe as it did elsewhere in the world.

The user expectation had been raised so high that a major disappointment was likely when people got a look at the technology, especially those who had wired web to compare it with. The WAP phones were too difficult to use, they were too slow, and they did not provide anywhere near the "Internet on a phone" experience touted in the ads. A crop of GSM phones outfitted with WAP required users to place a call to a web portal each time to send and receive data. Wireless carriers' data-connection plans were extremely expensive, and average consumers found it next to impossible to find WAP websites that worked on their mobile phones.

Only a modest number of wireless portals became available to support WAP applications, and there were some genuine constraints in redoing the web pages for WML. Sluggish performance, dropping connections and paltry content turned users off in droves, and they logged off for good. The much vaunted but now derided WAP technology could only attract 7 percent of mobile phone users a year after its launch. The WAP fiasco also brought new questions about Europe's credentials to do the Internet and the future roadmap to 3G wireless. Industry analysts also pointed to the fact that wireless companies, while copying the wireline model, were not used to data services mastered by the Internet service providers (ISPs).

i-MODE VERSUS WAP

The Internet-everywhere buzz in Europe came down to the reality of the 1-inch screen. The critics of WAP said that tiny screens couldn't hold enough information to be useful in most applications, and the numeric keypad was an abomination as an input device for controlling any amount of complexity. Moreover, the bandwidth was so minuscule that even small amounts of data took forever to transfer. But the i-mode experience had made one thing clear: a great deal could be achieved with tiny displays, small keypads, and low data rates. After a year of its launch, there were less than 2 million WAP users in Europe, a far cry from 13 million subscribers DoCoMo's i-mode had won in the same period.

While many observers reasoned that DoCoMo's mobile Internet success was a distinctly Japanese phenomenon, due in part to Japan's relatively low penetration of home PCs, i-mode's hitting of the sweet spot was not that shallow. The success of i-mode had proved that fun new services could attract a big market. DoCoMo had managed to create a vibrant mobile Internet experience largely by staying clear of the content business: it merely provided the platform and opened it to the web entrepreneurs, who enticed people with applications in thousands. Another important lesson: getting hung up on technology and not thinking about the service a company was delivering could be disastrous. DoCoMo aimed at fulfilling user expectations; the company didn't even promote i-mode as a wireless Internet service.

The "i" stood for information, not the Internet. "WAP failed in Europe because operators concentrated too much on the technology rather than the content," said Keiichi Enoki, who led i-mode project team. "It's like worrying about the quality of television sets before you have any programs." The i-mode service did have technical advantages, too. Unlike WAP phones, which required a slow dial-up connection to a server, i-mode offered an always-on link to the Internet. Another reason WAP failed while i-mode succeeded was the fact that DoCoMo prefixed hardware platform. That facilitated better display characteristics due to standard hardware modules in handsets. Furthermore, it was a much smoother shift to cHTML, the compact version of the web lingua franca, than making a wholehearted transition to WML.

Now the signs of what might be called "i-mode envy" were seen everywhere in the wireless world. There was a possibility that Western excitement about mobile Internet services could eventually find refuge in a number of i-mode myths. The i-mode showdown had brought a Japanese player to the offense on the global communications front for the first time. Now DoCoMo began aggressively promoting the service as a standard for mobile Internet appliances through multibillion acquisition deals in Asia, Europe, and America. For businesses outside Japan, however, achieving i-mode experience was not simply a matter of carbon copying a successful concept, but of attaining a greater understanding of the driving forces behind i-mode.

Despite all its exuberance, i-mode was still a proprietary system. And unlike the Japanese market, where one operator, NTT DoCoMo, defined standards that breathed life into i-mode, wireless companies elsewhere seemed to be unable to agree on a single standard.

On the other hand, WAP, an early attempt to reproduce the web for mobile phones, was a chic project of its time. It failed because its backers inflated the user expectations to a level almost equivalent to the Netscape-browsing Internet. However, as the stories of an impoverished user experience came out of Europe, the technology handily became subject to headlines like "WAP is crap." The growing epidemic of "WAPlash" struck fear among enthusiasts as the wireless web initiative was blamed for the root of all mobile ills. The whole idea of wireless web was exhilarating, but it also brought enormous engineering and business challenges.

Did WAP stand a second chance to engineer its resurrection? The WAP Forum thought so. Despite its failure in the first go, many experts believed all WAP needed was bigger transmission pipes and better consumer education. The market would grow rapidly, they argued, once 2.5G services such as GPRS and subsequently 3G became widely available, helping to make WAP connections many times faster. The WAP proponents were firm in their belief that this web-on-the-go initiative wouldn't join the ranks of failed technologies, no matter how good story trade press made. The truth of the matter, as always, was probably somewhere in the middle: WAP was at the initial stage of the value-adding ladder.

The idea of wireless web was just another stage in the evolution of consumer behavior that was taking shape since cellular phones kicked off in the early 1980s. The Internet we all knew was a smash hit on the PC but could easily flop on the handset. It was because the wired web was built for a different class of users. Using the same path to track down information via the phone proved to be a torture, so the phone makers needed to construct a different Internet experience altogether, one that would focus on mobility and would send services directly to the phone with a minimum of clicks. That was the moral of the WAP versus i-mode story.

REALITY CHECK

The mobile Internet had promised to shape into a full-fledged industry. The idea of Internet in one's pocket spread like wildfire, but once the dream turned sour, marketers were no more talking about the mobile Internet. The mobile Internet catch phrase sounded tedious by the end of year 2000. The construction of wireless Internet infrastructure was way behind schedule, and there was no clear business strategy in place for mobile data services. In the wireless industry at large, no one seemed to have the faintest idea what form the new data services would take, which services potential customers would buy, or even how to market the new products. Seldom had so many seemingly shrewd executives gotten their signals so crossed. The wireless revolution literally went on hold.

So what went wrong? For a start, wireless networks the world over were already struggling to accommodate the surging volume of voice traffic, so they could deal with data only at a snail pace. Moreover, the problems of dropped calls and limited coverage, which were already frustrating for the voice users, could prove devastating for data services. It was one thing to drop a voice call, but it was another thing when a user was in the middle of a stock trade or banking transaction. Combining the Internet with mobile phones posed business, technical, and cultural challenges.

The Internet users expected things to be free, and were prepared to accept a certain degree of technological imperfection. Mobile phone users, on the other hand, were accustomed to paying but expected a far higher level of service and reliability in return. Another question mark was why people would move to a wireless data platform when fixed-line Internet access was so cheap, and in some cases almost free. Other doubts surfaced as the wireless industry entered the uncharted waters of mobile data. How would data contribute to the bottom line—profits? How would a user want to pay for the data services? Then there were technology issues.

The notion of mobile Internet as the offspring of two spectacularly unpredictable technologies—computing and telecommunications—cut both ways because the two worlds represented starkly different visions of the wireless Internet. Those on the Internet side of the fence complained that wireless companies didn't really understand data networking; those on the wireless side grumbled that the Internet technology was flaky. To lay claim on the mobile Internet gold, wireless telecom industry was stretching far from its core market. Marketplace diversity aside, the path to a wider take-up of the mobile web was never going to be an easy one.

These were still early days for the mobile net: people tended to think of it as the Internet without wires rather than as something entirely new. Wireless access to the Internet, however, would almost certainly be different from the wired Internet experience. An obvious weakness, as mentioned earlier, was that the mobile net campaigners put too much emphasis on technology rather than making it beneficial to users. The wireless industry should have been quicker to pick up clues from the way earlier telephone technology had become popular, and more recently from the text-messaging conquest. They could contemplate, for instance, that success of non-voice applications involved setting up the right environment to allow such services to succeed.

In the hindsight, the phenomenal success of i-mode and SMS had built a foundation for the future success of smartphones. Text messaging had achieved astonishing growth at a time when the mobile phone industry was trying to dictate the deployment of WAP. Subsequently, though WAP had failed to grab consumer imagination, a segment of the wireless industry hoped it was the fault of the implementation rather than inherent resistance to mobile data. Here, they were taking some comfort from the popularity of text messaging and the roaring success of i-mode service in Japan. In a perfect irony, the success of SMS had left some industry experts with a new puzzle: whether it was pointing to a huge marketing demand for mobile net services or what most people wanted from mobile data was services that were simple, cheap and more like e-mail than the web. After all, text messaging made the best of the tiny handset screen while WAP failed to compensate the same small screen.

The Europeans claimed to have learned valuable lessons from the WAP failure. Their mantra now was selling applications, not technology, and the most triumphant application by far was text messaging. They were keen to point out that WAP hadn't gone away; it just branched into more practical applications, such as multimedia messaging service (MMS), the descendant of popular SMS.

The MMS technology with picture messaging at the heart of it was essentially WAP: MMS made use of WAP components for implementing transport and push mechanisms. After the WAP fiasco, wireless industry in Europe borrowed a page from the WAP book and rallied behind MMS, which made it possible to zap photos and graphics to and from handsets, just like the way text messages were sent in droves. MMS was not a giant technological leap forward, but a feasible feature made available on the handsets equipped with built-in cameras. The ability to take pictures with a handset and send them to other people was something that people wanted and were willing to pay for. Phones with color screens had also provided a strong impetus for snap-happy users.

MMS—which allowed the sending of still and moving images, audio, and text—was expected to become a springboard for mobile entertainment services of the future. It accelerated the adoption of digital photography and became a stepping stone to more advanced multimedia services that could become the staple of the coming 3G networks. The battered mobile phone industry hoped there would be more message-based entertainment to come, especially when advanced messaging services featuring color, moving images, and audio were rolled out. MMS had evolved from text messaging, a runaway success, and because MMS was based on WAP standards, it was possible to add a plethora of new features to it.

However, as it turned out, MMS was no miracle. The flipside was that MMS handsets and infrastructure from different vendors didn't work together smoothly, so users found it more convenient to stick with plain, old text messaging. MMS enhanced the consumer's experience of messaging, but in itself, it was merely an enabling technology. So the expectation of MMS to replace text messaging eventually faded; MMS was more likely to be used on special occasions like birthday greetings or photos from holidays instead of postcards. Again, the industry's great hopes for advanced mobile messaging were not fulfilled. The telecom industry was notorious for hyping new technologies, only to find out difficulties and delays in their development later.

Nevertheless, the transition from SMS to MMS was critical in creating new revenue margins per message for mobile operators. At the same time, wireless companies needed to focus on a new Internet, the one with services dedicated toward people on the move. Fortunately, DoCoMo had provided a role model in the form of its celebrated i-mode service. The first-ever network service linking mobile phone users to the Internet, i-mode was a ringing example of how to hook customers onto a new technology and turn a fat profit.

GREAT IDEA, SHAKY START

The unambiguous message the industry could draw from the failure of WAP as well as the success of text messaging and i-mode was that people would go for wireless data services if they were useful and inexpensive. The success stories also showed how important it was for telecommunications companies to innovate. With the ever-increasing potential of wireless data services, mobile phone operators needed to come up with creative new strategies to attract users. People wanted their mobile phones small because they looked cool and screens large so they could easily access the Internet. And the quality of wireless data would need to be as good as voice to come up to the user expectations.

Jorma Ollila, Nokia's chief executive during the early 2000s, acknowledged the challenges ahead, calling it "a big paradigm shift." Data was new, and that changed everything. Mobile phone companies were first taken off guard by the surge in voice services; they were surprised even more by the success of prepaid phones and text messaging. While struggling to match demand, wireless operators could do no wrong as very little marketing was needed for a voice-centric environment. Next-generation mobile services, particularly data ones, were far less likely to be a success by a similar fluke.

Beyond this mobile maze was a remarkable unanimity on the belief that the combination of wireless and the Internet would lead to a new revolution that could dwarf the one initiated by personal computing. Although we didn't see an explosion in wireless data initially, a fundamental shift was still quite evident. The sheer volume of SMS proved there was an appetite for mobile data services, and Japan had shown the world what could be termed as a crystal ball for the future of the mobile Internet. The mobile data market, which looked like a poor stepchild of wireless voice in its early years, began making significant strides once the notion of personal communications came into a holding pattern.

Clearly, there were all these pieces of action happening in relative isolation. Some industry analysts owed i-mode miracle to Japan's growing band of data junkies who spent most of their time outside homes and offices—on commuter trains. And according to them, text messaging was only popular because it was remarkably cheap. But then what could be said about South Korea, where wireless web services began to thrive as data-capable handsets became affordable by 2002. Over there, it didn't happen just because the content was attractive right from the start, but also because of the availability of slick new handsets for whom downloading games, cartoons, and richer-sounding musical ringtones were made so easy.

The real issue was not i-mode versus WAP, but the realization of new data services. A lack of applications, for instance, was a big complaint from consumers during the launch of WAP. Wireless handset makers, newcomers in the software territory, rushed to sell steroid-injected phones while ignoring the potential trouble spots. Consequently, an embarrassing episode of software glitches raised questions about the software acumen of mobile phone companies. Wireless industry probably believed in a more simplistic notion of the mobile Internet: build a network platform and let all sorts of capabilities or applications grow on it. Just like when Bob Metcalfe built the Ethernet, all he was looking for was to link a few computers to a printer at the end of the hall at Xerox PARC.

But such simplistic notions were in a stark contrast to the futuristic visions that WAP backers instilled in the minds of the consumers. In fact, WAP became a case study on the dangers of overhyping the technology. Despite all the debacles, however, the mobile Internet offered some tangible benefits by making mobile devices connected anywhere, anytime. So the promise of wireless web was not lost despite enormous engineering and commercial challenges.

In many ways, the mobile Internet during the mid-2000s was at the same stage of development as the wired Internet was in 1995. Nobody really knew which technologies or business models would win or what consumers or corporate users wanted. The only difference was about expectations. Some people wrongly expected the mobile Internet to be the same as the wired version, only mobile. Here, it'd be worthwhile to remember that the Internet had been around for more than fifteen years before it became usable on a wide scale. Scroll back to the early days of the Internet when web 2.0 wasn't there, and most people used CompuServe, AOL, or Prodigy—all on dial-up connections—and endured limited content and network capabilities, browsers with primarily text-based services, no graphics, and a few bits of information posted by even fewer people.

The wireless industry had to start somewhere. What mobile Internet went through over the period of a decade after its inception was somewhat reminiscent to wired Internet's journey into the commercial realm. The notion of wireless Internet subsequently turned into a made-for-mobile information highway where one could get on from nearly anywhere, and it'd continue moving in faster speeds. The next chapter encompasses the coming-of-age story of the mobile Internet sparked by Apple's celebrated entry into the mobile arena in 2007. As popular as i-mode was, it was nothing compared to the next i-product that would come along. The mobile Internet—after its humble beginnings in the late 1990s, its growing pains, and its misses and hits—was finally ready for the prime time. The mobile Internet 2.0 also brought to fore full-fledged browsers, pushing WAP-based micro-browsers into the annals of technology history.

The WAP pioneer Alain Rossmann was part of the legendary team at Apple that built the original Macintosh computer. He was also there when Silicon Valley launched the pen computing movement during the 1990s. The French entrepreneur faced skepticism when he began work on bringing the Internet on tiny mobile phone screens. When Rossmann told a wireless operator that people would eventually be using their mobile phones for e-mail, their answer was, "Why would you want to do that?"
Image source: SFGate

Keiichi Enoki—also known as Mr. i-mode—is considered the main architect of Japan's greatest consumer success since the Walkman. He was the one who hired both Takeshi Natsuno and Mari Matsunaga, the other two crucial figures in the making of i-mode mobile Internet service.
Photo courtesy of NEC Corp.

Takeshi Natsuno was the mastermind behind i-mode business model and multimedia-related services. The former Internet startup manager had joined NTT DoCoMo in 1997 just before his former employer, Hypernet, Japan's one of the first free ISPs, went out of business.

When Mari Matsunaga, former editor-in-chief of a classified-ad magazine, joined the i-mode project in 1997, she had never used the Internet and hated cell phones because she thought people using them in public were rude. Matsunaga led i-mode's content strategy while inking deals with third-party mobile websites. As the content head, she insisted that i-mode would not pay or charge any content providers except for billing services, giving them an incentive to stay in touch with their users and to improve their services. Matsunaga, who fought ferocious battles with the engineers during the project, strongly believed that users didn't care about technology and that they just wanted usefulness.
Photo credit: NEC Corp.

MOBILE INTERNET 2.0

"Finally, someone has delivered a cell phone with a compelling web experience."
— Andy Bechtolsheim, Sun Microsystems co-founder, commenting about the iPhone in 2008

When Steve Jobs revealed the iPhone at the Macworld Conference and Expo in San Francisco in January 2007, hardly anyone anticipated that this sleek new gadget would set off the mass adoption of handheld Internet access. The iPhone turned out to be the breakthrough Internet communications device that the wireless industry had been longing for since a decade. Until now, mobile phones ran on disparate pieces of software, had less memory, and operated under the constraints of pay-per-byte wireless networks. Consequently, the mobile

Internet we had was a mere stripped-down version of the real thing. Apple understood better than anyone that instead of leisurely browsing and searching like on a PC, the mobile Internet would take off by focusing on sending and receiving timely, relevant nuggets of information.

The way i-mode created the first mobile Internet success story by balancing technology excellence with content innovation in fact provided a classical case study for future projects. The phenomenal success of i-mode boiled down to NTT DoCoMo's ability to understand the constraints so well and then make the best of what was available at that time. And fashion in which DoCoMo conceived a functioning ecosystem for i-mode to succeed by letting small outfits run the content show was also quite remarkable. In retrospect, it would be safe to assume that Jobs and company learned all the right lessons from the success of i-mode and the failure of WAP. Because that's exactly what we witness in the mobile Internet's renaissance set off by the Cupertino, California–based company a decade later.

Probably better than any other player in the market, Apple understood that one of the biggest draws for smartphones was their ability to access the Internet without having to find a PC. Because handsets with small screens unsuited to Internet browsing, it took Apple's fanatical attention to usability to bring Internet-connected devices to the masses. Apple borrowed the PDA's large-screen prerogative and the use of icons from personal computers and inspired the next-generation form factor for mobile phones. Moreover, Jobs and his comrades knew very well that GSM pipes were too narrow, and accompanying wireless modems were too cumbersome, so they waited for the fat pipes that could transmit web pages, pictures, and other data at higher speeds.

But probably the most critical lesson that Apple seemed to have fathomed was that the software obligation was growing in the smartphone realm. So, first, Apple carefully calibrated its mobile Internet offering with the introduction of Safari, a graphical web browser, which was also the native browser for the iPhone iOS software. It was the first mobile browser to display mobile web pages identical to those displayed on a desktop computer without sacrificing usability. Previous mobile web browsers had been unable to capture web pages in the same format, often resorting to limited web viewing or incorrect formatting. Apple had developed the original Safari browser as part of the Mac OS X operating system.

Another crucial lesson that Apple must have learned from the WAP debacle was that the browser was not end-all, be-all in the mobile Internet game. WAP was, in fact, a crude form of web browsing on mobile phones. So to make the Internet truly useful, the iPhone would rely on specialized applications called widgets, which combined device data on information such as the current location with online data and services. These small applications ran on mobile devices and accessed information on the back-end. That way mobility brought new aspects to the

Internet, enabling cellular handsets to offer an Internet experience truly comparable to the broadband access that people had at their personal computers. Widgets or apps became the key enabler for the Internet-on-the-go.

In the quest for the riches of smartphone, mobile Internet was the starting point. The notion of mobile Internet had long been an article of faith among the optimists in the wireless world. To lay claim on mobile Internet, however, the wireless industry had to stretch far from its telecom roots. It was a fascinating idea on paper, but as we saw in the preceding chapter, mobility wouldn't come without a fight. Now, after the ascending of the iPhone, the coalescence of mobility and the power of the Internet that visionaries had been dreaming for a decade suddenly became a reality. The humble radio was finally poised for the Internet touch on the mass market. And Apple's marvel of buzz marketing in the making of the iPhone archetype was somewhat reminiscent of DoCoMo's i-mode marketing triumph.

The technology press in the late 1990s was filled with predictions that most of the Internet activities would occur on mobile devices. And many people became agitated when the bonanza didn't occur within a year or two. The prediction had finally come to pass, thanks to the iPhone. Now, in the post-iPhone wireless world, the mobile Internet market was taking off even faster than predicted. These were the years in the late 2000s when mobile Internet adoption actually happened eight times faster than web adoption on the PC during the mid-1990s. No wonder Apple quickly began grabbing the market share. At this junction, when Apple was redefining the mobile handset through a knock-out web experience, industry watchers began to wonder if anybody could stop the Silicon Valley computing icon. Enter Android!

GOOGLE'S SECOND ACT

Android enabled handset makers like Motorola and Samsung to develop credible rivals to the iPhone, and when these companies started gaining traction, Apple's momentum began to slow down. But why would Google create its own mobile phone platform in the first place? The simple answer was that the Internet dynamo—which gave away a ton of services like Gmail, Google Docs, Google Reader, etc., and in turn, raked in tons of advertising revenue because of these and other services—wanted to take these services even further. There was strong evidence to suggest that in the search and advertising domains users were likely doing fifty times more data queries with a smartphone than on a basic wired connection. So the future of Google would be influenced through mobile devices one way or the other. And Google wanted to ensure that it positioned itself for the mobile future with the development of its Android mobile operating system.

Initially, Android looked less about money and more about iPhone disruption. However, the driving force behind Android, according to Google founders, was a vision to produce phones that were Internet-enabled and featured good browsers because that didn't exist in the marketplace at the time of Android's entry into the smartphone fray, which more or less coincided with the launch of the iPhone. In a nutshell, it was imperative for Google to translate its web dominance into the mobile arena if it wanted to remain relevant. Google, lacking Apple's consumer market know-how, had its own struggles in adapting to the new world of smart devices. However, while Apple's overwhelming influence on the media world was probably its biggest strength in the mobile Internet realm, Google's edge was better technology. Android's secret weapon was really the network effect.

The Apple-Google rivalry wasn't a mere squabble; it came down to the future of the web and subsequently Google's place in it. The industry was on the verge of a mobile revolution—riding on the back of smartphones—and the biggest wave ever was to hit the world of computing. Just as mainframes gave way to minicomputers, which in turn gave way to personal computers, the PC was now being seen as displaced by smartphones and tablets. The smartphone users were downloading services and apps in large numbers and every device sold was to generate an ongoing revenue stream. The leading web analyst Mary Meeker forecasted in 2010 that, by 2013, the mobile Internet ecosystem—capital spent on access fees, mobile-commerce, paid services, and advertising—would be worth more than half a trillion dollars per year.

By 2011, the industry was witnessing super-smartphones in the works with much larger on-board memory and faster multicore processors. If trend lines were extrapolated, smartphones would most likely be the conduits for a majority of people to access the Internet in the coming years. It was that threshold point that eventually reshaped the smartphone into the ultimate gadget of the twenty-first century. From education to healthcare, energy to battlefield, there was hardly a discipline that didn't have smartphones and mobile Internet on its product roadmaps. Not surprisingly, therefore, it was a battle for literally every person on the planet and that's why companies like Apple and Google saw this market as worth fighting for. Both Apple and Google were headed by recognizable visionaries like Steve Jobs and Eric Schmidt at the time of mobile Internet take-off.

Google discovered the crucial reality just in time that it needed to find real success in the new world of the mobile Internet and become part of the next major evolution of the web. For Google, the desktop metaphor was fading. If phones were going to replace PCs as the main gateway to the Internet, why would consumers tether themselves to a PC, especially when phones were growing more and more powerful and becoming cheaper? That obviously made the online advertising giant very keen for dominating the mobile web experience the same way it ruled the desktop Internet domain.

There was another intriguing dimension of the iPhone versus Android fray. Both handset manufacturers and wireless network operators were increasingly becoming commoditized, and they were desperate to find new sources of revenue. Between them, the most valuable thing they had was control over what went on the phone before it reached the consumer: content and applications. That was exactly what Apple and Google wanted to control as the future of wireless and computer industries shifted toward the mobile Internet. Google, especially, had a lot at stake. Google had to pay Apple to keep Google as the default search engine on the iPhone. So if other major handset makers would auction off the default search services on the phones they shipped, Google might have no choice but to buy their support and it surely wouldn't come cheap. So in the endgame, Google needed handset makers as allies as much as they needed its free Android software.

Inevitably, Apple and Google represented two somewhat colliding visions of the mobile Internet. Apple, through its iconic offerings like the iPhone and the iPad, was generally perceived as creating a "walled garden" where content and games needed to be approved by Apple and accessed through its App Store. Putting such preconditions on which technologies Apple would support on its devices was increasingly seen as akin to trying to control how the mobile Internet developed. For instance, Apple saying yes to HTML5 but no to Adobe's Flash. Google's business model, on the other hand, was built on the openness and the anarchy of the Internet. The Internet search kingpin made money on the web that was completely open, so any control on the content of the Internet could threaten its revenue stream.

The smartphone turf now turned into a fight for the control of the mobile Internet market, where all digital convergence was headed, where the Internet was headed, where media was headed, where computers were headed, and where digital money was headed. Although people generally thought of next-generation gadgets sporting the mobile Internet as a phone, as established in the previous chapters, in fact, they were a computer. And we learned that, for computers in the new ubiquitous era, it was imperative to have a complete and functional connectivity. Ironically, at the time of the mobile Internet take-off, when cellular communication was being invented all over again, Nokia and Symbian software were nowhere to be seen. And Microsoft, who had a habit of being late to the web party, would again be playing catch-up in the mobile Internet space.

LOST IN TRANSITION

Mobile telephony allowed us to talk on the move while the Internet turned raw data into exciting services that people found easy to use. To make the best of both worlds, smartphones kicked off the transition from voice-centric business model, and that was followed by mobile phone industry's rush toward data applications. As a result, these two technologies began converging to create a new communications nirvana they called mobile Internet. The convergence of the two fastest-growing communications technologies of all time—mobile phones and the Internet—would make possible all kinds of new services and create a vast new market as consumers around the world started logging on from their Internet-capable phones.

For a start, mobile phone operators tailored a variety of online services according to perceived users' demand and accomplished a modest success. Users could check bank balances, weather updates, traffic reports, and news headlines either through SMS or by using a clunky Internet connection. Other notable initiatives in those early days included web media giant Yahoo! collaborating with cellular operators and Internet pioneer America Online jumping the bandwagon to bring instant messaging service to mobile phones. The notion of the mobile web kept crawling on the back of rather primitive smartphones, which had begun providing a reasonably good net connection by 2002. Meanwhile, the size of displays on cell phones kept on getting bigger in order to accommodate more data services.

Here, it'd be worthwhile to note that PDA makers—the descendants from the computer industry who eventually turned into phone wannabes—were the ones coming up with mildly successful mobile Internet calls. Inevitably, these devices looked more like PDAs than phones, and therefore, these smartphones in the early going were mostly targeted at high-end business users. Research In Motion (RIM)—which later renamed itself as BlackBerry, its best-known product—took an early lead when it added phone and other features to the wireless adaptation of Internet's first killer application: e-mail.

At around the same time, Handspring gave birth to yet another smartphone when it enhanced its flip-style Treo 90 PDA by adding dual-band GSM capabilities. The Handspring Treo, which later became Palm Treo, drew many customers with its winningly innovative design and got an early traction in the upper end of the phone market. With organizer, phone, and Internet features all on hand in a single, compact device, the Treo smartphone was a marvel of thoughtful design. The Treo featured options for connecting to the Internet on a descent-sized color screen along with a tiny but full keyboard. It was a terrific product for both phone-centric and data-centric people who wanted it all.

Another notable product that symbolized mobile Internet's awkward adolescence was Danger Inc.'s Hiptop phone, which T-Mobile marketed in the United States as Sidekick. Hiptop, which could be used to surf the web and to send and receive e-mails, was the first phone to offer both a full QWERTY keyboard and a built-in version of America Online's popular instant-messaging service. It quickly became a status symbol among the rich and famous in the United States in those early days of the mobile Internet.

The early stage of the mobile Internet development was mired by handsets with screens that were too small and had low resolution; their keypads were diminutive and they couldn't display all formats on the web. Then came the mobile Internet's defining moment in summer 2007 when Apple launched the iPhone with a clever design and some nifty capabilities, and blew up the wireless industry. Apple meticulously worked out these challenges, and that's how the iPhone became an inflection point for smartphones. The vision of multimedia wireless Internet had finally come true through the iPhone's beautiful user experience. The mobile Internet had embarked on the next phase of its makeover.

Until this tipping point, the notions of wireless Internet and smartphones went hand in hand, galvanized by high-profile initiatives such as WAP, Symbian, and Bluetooth. Now the wireless industry had finally found the silver bullet it had been longing for many years and the ultimate contest for the biggest prize had begun. It was the most dynamic time in the brief history of smartphones. As Internet moved from desktop to pocket, the nature of mobile phone usage was going to go through a profound change.

The mobile Internet technology had opened an entire new frontier for newcomers like Apple and Google. For the entrenched players, however, it was all becoming too much. Palm Inc.—a smartphone pioneer and the company that launching the PDA craze—eventually went up for sale amid financial woes and was acquired by Hewlett-Packard in spring 2010. However, a far bigger shake-up was coming to the door of BlackBerry, the maker of the first-ever successful smartphone. BlackBerry was growing wildly at the time Apple rewrote the rules of the mobile Internet. The famously addictive device popularized e-mail on-the-go and turned its Canadian maker into a US$34-billion company and the business world's leading supplier of smartphones.

BlackBerry was everywhere: boardrooms, restaurants, and kitchen tables. The device that earned the enduring nickname of CrackBerry represented more than half of all corporate smartphone users in the United States.

The dark little devices would vibrate every time a new e-mail arrived, delivering a tiny thrill that millions of employees came to both loathe and desire. However, BlackBerry didn't take the web experience as seriously as it should have, and eventually it began to look like a tired remnant of yesterday's technology. BlackBerry's weak browser capability became a huge issue leading to significant erosion in its customer base. Later in 2010, three years after the iPhone launch, BlackBerry completely revamped its browser; it was much faster, toggled more easily between windows, and was better at handling RSS feeds. By this time, however, BlackBerry had lost a significant market share to snazzier rivals such as the iPhone. In September 2010, three years after Apple delivered the first compelling mobile Internet experience, the iPhone surpassed the BlackBerry in quarterly sales with 14.1 million devices sold, compared with 12.1 million for BlackBerry.

The failure to see the rise of the mobile Internet in the commercial arena and to grasp it in a timely fashion had put BlackBerry into a transitional moment. Industry analysts were overwhelmed with concerns about the future health of its mobile platform. However, the disruption caused by the second coming of the mobile Internet didn't only mark the end of the BlackBerry elite; it also became a predicament for Nokia, which had once dreamed of reshaping the Internet. Nokia had been vocal and ambitious about its role in the Internet world. In fact, Nokia firmly believed that no other company was better placed to be the standard-bearer of what its marketing people endlessly called the "mobile information society."

No one had embraced the idea of wireless web more enthusiastically than Nokia. "In ten years time, I would like Nokia to be dubbed as the company that brought mobility and Internet together," said its CEO Jorma Ollila back in 2000. "It's not going to be easy, but this organization loves discontinuity; we can jump on it and adapt. Finns live in a cold climate: we have to be adaptable to survive." Now the fact that Nokia, though still the world's number one smartphone manufacturer in 2011, had stumbled several times in its efforts to catch up on the mobile Internet was a harbinger of the coming shake-up. The Finnish mobile phone maker was failing to fight back the onslaught of Internet-enabled phones like the iPhone. It was ironic that the company that got the whole smartphone thing going now found itself at a peculiar crossroads. What it really needed to do was deliver the promised land of the mobile Internet, and only then, this Finnish wonder could regain its star status.

MEDIA AT YOUR HAND

In 2010, Morgan Stanley projected that in five years the number of users accessing the Internet from mobile devices will surpass the number of users who access it from personal computers. Mobile phones were now the focal point of all forms of digital convergence. The Internet kingpins like Google and Yahoo! were saying that the future of the Internet was on mobile phone. The PC behemoths such as Dell and HP asserted that the future of computing was on mobile phones. The media giants from TV to music to video games were promising to make all future media content available on mobile phones. With the advent of smartphones, and by that extension, of mobile Internet, it became evident that once-dreamy notion of convergence between telecom and computer worlds was for real, and it also brought forward another intrinsic element of the smartphone anatomy: media.

The wireless industry had been hung up on the soap opera-like marriage of mobile phones and PDAs for so long that it took its eyes off the real story: digital content. If the mobile Internet would be the starting point of the smartphone revolution, then its architects should have better known that the Internet was all about the open and free flow of information. Eventually, when the real mobile Internet took off under the wings of the iPhone, the wireless industry woke up to the battle over the control of digital content. Likewise, media industries across the spectrum answered the mobile call only after Steve Jobs reinvented smartphone at the crosshairs of the mobile, Internet, and media businesses and gave smartphones a business model that actually worked. After that, all these pieces from both sides of the fence began falling in place, leading to a whirlpool that would turn both wireless and media worlds upside down much faster than anyone ever could have imagined.

In retrospect, however, the media ascendance in the mobile fray was not that sudden. In 1998, for instance, one of the first success stories of selling media content via the mobile phone came through the sale of ringtones by the Finnish mobile operator Radiolinja. Soon afterward, other media content appeared on then-tiny mobile phone screens, including news, videogames, jokes, horoscopes, television clips, and advertising. Most of the early content for mobile phones tended to be copies of legacy media, such as the banner advertisement or the news-highlight video clips. That was followed by unique content for mobile phones, ranging from ringtones and ring-back tones in music to video content, exclusively produced for mobile phones.

Here, given the benefit of hindsight, it could be safely said that the European WAP effort for the riches of the mobile Internet partly failed because it concentrated too much on the technology and too little on the content. Mobile phone operators were reluctant to share revenue with content providers. They failed to see that data wouldn't catch on until every party to the transaction made money. That's something NTT DoCoMo recognized early on. The Japanese firm established a model that provided revenue for every player in the information value chain.

Eventually, the mobile phone got itself acknowledged as the third screen after TV and PC; it was also called the "seventh of the mass media" with print, recordings, cinema, radio, TV, and the Internet being the first six. In 2006, the total value of mobile phone-paid media content exceeded Internet-paid media content and was worth US$31 billion. Next year, the value of music-on-phones revenues equaled US$9.3 billion and mobile gaming revenues were worth over US$5 billion. The evolving processing power and robust software capabilities of smartphones radically enhanced content richness and helped create fantastic audiovisual effects. Available apps ran the gamut of interests that included games, news and weather, maps and navigation, social networking, and music.

For instance, if something would happen in the news, mobile users went right to their *The New York Times* or *The Wall Street Journal* app. There was a clear evidence that consumers were quickly switching away from wired web pages and toward apps as their preferred way to access the Internet. Smartphones were now able to access all of the user's content and media, including work and personal documents, music, video, and so on. Some of this would be stored on the device itself; some would be stored in the network cloud. Smartphones and mobile Internet, therefore, created a revolutionary new distribution channel, which represented the next big opportunity for the media.

With Internet on-board, the smartphone could be a person's content-consuming device—with this screen, users could watch a movie. It could also be a user's reading device—the screen was big enough for people to have a reasonably good reading experience. And all of user's reading content could go with him or her anywhere. Having access to content and being able to display and transfer large amounts of information would let people do more with their devices. That included watching popular TV shows, listening to music, shooting and storing high-quality pictures and video, and sharing it via tools such as Flickr, Facebook, and YouTube.

Besides gaming, the next most profitable commercial apps were the ones that connected users with valued content for which they were willing to pay extra; for instance, apps for showing live sports. Next up, after the initial, low-cost downloads, annual user fees could exceed US$100 for apps like the Netflix and Hulu mobile video services. The thousands of apps built around platforms like Android and iPhone, mostly by third-party developers, encompassed almost every area. Meanwhile, the mobile Internet experience continued to optimize and innovate. Case in point: a year after the launch of the iPhone, 3G version was added to improve the browsing speeds. The actual look and feel of the handset varied only slightly; most of the changes in this release were inside the iPhone.

A NEW ERA OF DIGITAL CONTENT

The above history of media's foray into the mobile world only entails to cell phones in general. The smartphone, on the other hand, wasn't just another type of handset but was a full-fledged computer. It came loaded with software and doubled as digital camera and portable entertainment center. That's why companies like Apple and Google didn't see themselves as selling mobile computers, but instead connected Internet devices—truly a sea change from the traditional PC business. Apple had built one of the most successful media businesses of the Internet age: iTunes, a content distributor. Google, apart from being known as the web gatekeeper, was undoubtedly the largest media empire capitalism had ever created. And finally—both companies wearing the hallmark of Silicon Valley innovation—understood the true value of an entire ecosystem.

There was another critical dimension hidden in the contrasting worlds of technology and media. On the technology side of the Internet, there had been a lack of profound understanding of media formats as technologists were often wary of any involvement with content. Conversely, on the media side, few people knew about technology. That led to disconnect between two fundamental building blocks of the mobile Internet fray. Steve Jobs, the Macintosh Man, had built two of the most successful media businesses of recent times: iTunes, a music content distributor, and Pixar, a movie studio. So he perfectly filled that void. But probably not even Jobs and Apple's senior executives would have foreseen either the speed or the depth of the transformation that would eventually take place at Apple and the industry at large.

Apple had a history in media business that went as far back as 1984 when it defined the fundamental elements of desktop publishing on the Macintosh platform. About two decades later, Apple's industry-changing music player iPod effectively killed the promising portable media player (PMP) market. The iTunes store, which arrived in 2003, allowed users to download music from the Internet for their iPods. However, by adding the Digital Rights Management (DRM) to its offerings, Apple forced the user to only use it for iPods. The iTunes was one of the most successful software packages in history, installed on more than 125 million computers worldwide and used for about 70 percent of all digital-music purchases as of end-2010.

Apple now sold TV shows, movies, games, and even university classes through iTunes. The company also used iTunes for a medium that could be its trump card: mobile apps. With iTunes being so well integrated and music players that seemed more like cultural icons, the iPhone came along as one of the best phones for music and general media capabilities. The iTunes having created a frictionless distribution system made getting an app on a consumer's phone as easy as getting a song on the iPod. The fact that apps became synonymous with music on iTunes was a harbinger of new opportunities for brands and media companies. First, onset of the iPhone transformed the mobile into a unique channel for media, and then in 2010, Apple's iPad sensation was aiming to redefine the magazine and newspaper industry.

Publishers had put a lot of faith in devices like the iPhone and the iPad. As traditional media struggled to adapt in an evolving Internet-centric landscape, and its audience dwindling, there was a strong drive on how this extraordinary mobile medium of smartphones and tablets could connect to audiences and make money. Still, the business model pioneered by the likes of the iPhone and the iPad was not without initial stumbling blocks. While publishers wanted their mobile content to be monetized with advertising, there was this issue of control over the medium as both publishers and companies like Apple wanted to drive the overall business process. Nevertheless, the advent of media on the smartphone platform was producing all kinds of new possibilities in content access and playback domains through browsers as well as native apps.

Smartphones with the power to access digital content could have a profound impact on people's lives. While mobile initially didn't have the volume of the wired Internet, it did have more engagement and targetability. The mobile Internet was also a lot less cluttered than its wired counterpart; there was one ad per page on mobile versus up to five competing for attention on the wired web. Moreover, the personal nature of the mobile Internet made the smartphone a lot more intimate as compared to the online and the TV experiences. For an Internet-enabled smartphone, it was the users' screen, they saw it every day, and that made a lot of difference.

The smartphone embodied a fusion for all sorts of portable devices—from camera and music player to gaming and navigation gizmos—and now that it had all been seamlessly connected to the Internet ecosystem of apps, services, and content, having a device that was always connected and was always with the user could radically shift the wireless paradigm. These new-age gadgets would bring mobile users the ability to store and retrieve mountains of information and perform tasks like navigating unfamiliar terrains. Smartphones equipped with cameras and Internet connections would be akin to mobile surveillance points and what would the world like with billions of such surveillance points.

The mobile Internet had finally unleashed the huge transformative potential that smartphone designers had envisaged, and from here, there was no looking back. Now that Apple had crystallized the concept of the web-ready phone, the question was where the mobile phone would go from here. On the outset, both wireless and media worlds were gradually coming to terms with swaying each other's territories in the new order facilitated by the mobile Internet. However, the media conundrum on the mobile screen was about to follow yet another showdown—a clash between two conflicting visions of the Internet in the hand.

The iPhone refused to rely on the "baby Internet"—as Jobs put it during his Macworld 2007 keynote address—and launched a compelling mobile web experience through its Safari web browser.
Image source: GadgetLite

Steve Jobs and Eric Schmidt led their respective companies to mobile Internet stardom. They are seen here sharing the stage during the early days of the iPhone. Their friendship would soon turn into a bitter rivalry.
Photo courtesy of *The Guardian*

4 THE INTERNET VERSUS THE WEB

"Even when the Internet was first starting out, computer users preferred apps over browser tools. The reason is speed; an app is much faster."
— Adam E. Ornstein, founder of VideoTagger software provider Electric Happiness

In 2007, when Steve Jobs demonstrated the iPhone to a stunned audience, Apple endorsed only a single way for having mobile Internet functionality onto the iPhone: via a web browser. Almost immediately, a jailbreaking community came into existence as developers—frustrated by some of the limitations of HTML and Safari mobile browser—created and distributed their own native applications. Apple didn't approve or support this, and that led to perceptions about Jobs' assertion that websites were more than enough. The term "native app" was used for applications designed specifically for a platform, in this case a mobile phone operating system like iOS.

There is strong evidence to suggest that, in the start, Apple didn't anticipate that native apps from third-party developers would become the single most-requested feature from consumers in the months following the iPhone's launch. The company had this skeptical view that consumers would only embrace web apps, making native apps little more than a curiosity. However, Apple was prepared to roll it out to have third-party developers on its side. If you read between the lines, the Apple press release dated June 11, 2007 had an apps blueprint written all over it. Moreover, according to some media reports, soon after the i-Phone's launch, Apple was already playing with a software development kit (SDK) for developers to create native apps for the iPhone.

A native app didn't need an Internet connection to run. It was much faster than loading a web page, and because a native app ran directly from the phone, it had easier access to hardware components like a microphone and camera. Apps were built around the same standard technologies created to build websites—primarily HTML, CSS and JavaScript—but an app did some tweaking of browsers and hardware to reach nearly every device in the world that ran a browser. Many apps were simply dedicated web browsers that used web-based standards and technologies to display and manage data—and some apps did that in a way that added a lot of value for users. However, the native apps people used on smartphones or tablets were tied to a specific operating system. An Android app could only run on a smartphone running the Android and an iOS app could only run on an iPhone or iPad.

Fundamentally, there were two things apps did well that the web did not: information processing and secure transactions. In a 2010 *EE Times* story titled "Mobile Internet apps moving e-commerce off the Web," R. Colin Johnson discussed in detail how apps offered a faster connection to the information that people most wanted and needed, and apps providers did that by cutting through the clutter of data. For mobile users, apps offered speed to information without having to mess with browsers, search engines, and URLs. As for security, with web pages, mobile users were subject to spoofing attacks plus all the vulnerabilities of the browser they were using. For example, data could be copied and pasted between windows.

Apple's iTunes led the way in demonstrating that apps could be locked down. Apple claimed a flawless security record despite having more than 150 million credit cards on file, processing 230,000 new iTunes activations per day, and serving a total of 120 million devices in the field in 2010. One of Apple's motivations in making iTunes an app vehicle was to have a consistent user experience, which would not be possible with all the different browsers on the web, but the second reason was the much cleaner security that could be provided with a proprietary apps platform.

Of course, apps were only as good as their authors, but when done well, they bypassed all the messy sifting required to mine the worthy information from the billions of web pages cluttered with banners, flashing animations and unwanted pop-up windows. Apart from the fact that native apps provided improvements in performance and storage, they could also work offline. Moreover, placement of apps on the phone's home screen made them more likely to drive repeated usage. In a nutshell, apps presented content providers a simpler experience that took better advantage of the phone's hardware features by exploiting the platform's built-in technology. Apps were programmed separately for each mobile platform, much as game code was tailored for specific gaming consoles, and as a result, apps fully utilized the special media-handling capabilities of each handset. However, as we'll see later in this chapter, this trait cut both ways.

By 2010, essentially most of the phone-based Internet usage was coming from Apple and Android mobile devices. Apple had launched a new wave of the mobile Internet growth on a platform that now largely bypassed the browser and many saw this apps revolution as spelling an end to Google's web domination. Because the handset screens were smaller, mobile traffic suited well for specialty software, mostly apps designed for a single purpose. For the sake of an optimized experience on mobile devices, users were willing to forgo the general-purpose browser. So they used the Internet, not the web.

Although Microsoft's Mobile Internet Explorer, Apple's mobile Safari, and Android's Chrome-lite were being refined over time, there were legacy issues that they inherited from the first-generation freewheeling Internet. For instance, while Apple's iPhone allowed users to browse any website that supported Safari, Microsoft's devices required that sites be built for its Mobile Internet Explorer. And the browser on Windows Mobile devices wasn't a standard browser, so it didn't work with a lot of websites. Not surprisingly, therefore, four out of five developers said that their users preferred native applications to mobile websites because of user experience expectations.

Unlike the wired Internet, native apps grew much faster on the mobile platform—up from virtually nothing in 2008 to more than 7 billion downloads in just two years. And that led to a new conflict for the riches of mobile Internet gold. The new mobile Internet now found itself riding on two parallel tracks: legacy browsers for the web-based environment and new-age apps pioneered by the iPhone.

The trade media and public at large tended to use "the Internet" and "the web" interchangeably for the same network. The web, however, was an application that ran on the Internet. But the web had also been the Internet's killer app, transforming the network of networks created by the U.S. Department of Defense for collaborative research into a global phenomenon having reached more than 2 billion users. Not surprisingly, therefore, the web, which was a major layer of the Internet, became synonymous with the Internet.

THE APP INTERNET

Apps culture was transforming web's commercial cravings to more trustworthy, easier-to-use services, thus reserving the openness of the web for its original purpose: the casual consumption of information, free from the security and privacy protocols and facilitated by a browser interface otherwise known to be chock-full of vulnerabilities. Mobile widgets did that by excising from the web and creating complex and security-conscious commercial applications.

Apple's adversary BlackBerry was initially critical of the ecosystem of applications for the iPad, the iPhone, and the iPod touch, saying that users didn't necessarily need an app while they had the web. The maker of BlackBerry gadgets acknowledged that there was a role for native apps, but was adamant that the web browser remained the best way of getting information on mobile devices. To bring the mobile to the Internet, it countered, users didn't need to go through some kind of software development kit; instead, people could simply use the existing web browsing environment. However, BlackBerry, like other smartphone makers, was forced to go the apps route in the quest to offer a viable mobile Internet experience to its customers.

Next, even though Google's preference was for services accessed directly over the web, rather than through pieces of specialized software, it was forced to follow suit as well. Smartphones upended the software world, and overnight, everyone wanted to build games and apps just because the iPhone had offered exciting new ways of using this technology. Now, apps were seen as essential in order to build and expand a thriving mobile ecosystem. The apps culture quickly spread to app stores for the Android, the BlackBerry and Symbian operating systems, creating a parallel universe of iTunes-like walled gardens that could grow to be just as expansive as, but much more secure than, the web.

Especially, Google needed to find real success in this new world where the Internet was evolving to suit mobile apps. Google boasted one of the best performing browsers on a mobile phone. According to Andy Rubin, the tech whiz who oversaw the Android empire, it was the fastest, and it was the smallest. And he vowed to add more functionality to the browser to give it an updated user experience. Evidently, Google wanted to get beyond the world of apps ushered in by Apple, toward a category of devices more suited to its own technology preference; ones that were built exclusively to channel web services.

However, if smartphones continued moving toward Apple-pioneered app-store environment that required unique development for each platform or mobile OS, web service models wouldn't remain viable anymore. Especially, when Apple had demonstrated a proven revenue model in which users were comfortable making app purchases. The so-called "app Internet" was the winner in part because of the continuing increase in computing power, both in the cloud, where giant server farms stored and processed a vast amount of data, and in the devices like iPhone and iPad. The magic of the apps economy also stemmed from a combination of distribution, monetization and add-on capabilities.

Native apps had evidently won the first round of the battle for the riches of the wireless Internet. Partly, that happened because mobile bandwidth had not kept up with these changes brought forward by smartphones, so the web gave way to a world of apps that processed and displayed the data coming from services in the cloud. Native apps were distributed through app stores controlled by the owners of the platforms while web apps were distributed through the open web and the link economy. Next up, native apps came with one-click purchase options built into mobile platforms while web apps tended to be monetized more through advertising, and payment platforms were apparently less user-friendly. Last but not least, native apps could do a lot more than web apps.

However, software developers couldn't afford to support every operating system. Apps were expensive to develop for a variety of platforms such as iOS, Android and Windows Phone. It was costly for small enterprises to develop apps for different platforms; they were forced to write Java for Android, Cocoa for iPhone, JavaScript for the browser, ActionScript for desktop, and so forth. Moreover, apps were difficult to build and maintain, and making money on them was a challenge in itself. So once the mobile OS space became increasingly crowded, mobile app developers with small staffs became far more interested in designing one-size-fits-all in-browser web-based apps that provided mobile users experience comparable to native apps. Take the case of Microsoft's Windows OS for desktop computers: the fact that it became a de facto standard made it easy and relatively cheap for software developers to produce PC apps and that had been a tremendous benefit to society since the mid-1990s.

So, in early 2010s, when Google engineers began advocating a new architecture that would blend the web and native apps, it came as a welcome relief almost across the board. Such a mash-up would enable developers to create an application for the mobile phone and then have the freedom to emulate it as a web application. In other words, if a standard web browser could act like an app, offering users a similarly clean interface and seamless interactivity, users perhaps would resist the trend to the paid, proprietary apps. For this very premise, the wireless web had its hopes set on an upcoming technology known as HTML5, designed to handle audio and video internally without the need for browser plug-ins like Adobe's Flash and Microsoft's Silverlight. It had gained a significant attention as a web-building code that had both Apple and Google behind it.

Apple, Mozilla Foundation and Opera Software had joined hands to create Web Hypertext Applications Technology Working Group (WHATWG) in 2004 and set out to develop HTML5. Two years later, World Wide Web Consortium (W3C) decided to stop work on XHTML standard and began collaborating with WHATWG to help evolve HTML5 as a full-fledge technology. In 2008, the first draft of HTML5 was introduced; however, the technology continued to evolve. Mozilla Firefox took steps to allow HTML5 to be viewed on the browser, and eventually, Chrome, Safari and Internet Explorer followed suit. Ian Hickson—who later joined Google—shepherded the standard for years as the editor of the HTML5 specification, first through the informal WHATWG forum and later with the more buttoned-down W3C.

HTML5 had little to do with HTML, web pages and links. In fact, it was an umbrella term for the tremendous advance in JavaScript capabilities in modern browser engines. HTML5 promised developers tools to build rich web-based apps that ran on any device via a standard web browser. So it promised to be the "genes" from which much of the next-generation web would spring to life: websites, content, and web-based apps could be partly or wholly coded with it.

One of HTML5's biggest sells was its flexibility. Regardless of what device a website or app showed up on—iPhone, Android tablet, desktop, laptop, all of which had decidedly different form factors—developers could use almost the exact same software code for each device platform. Therefore, it would cut down on the time, effort and manpower required by mobile app developers to program native apps in the crowded mobile OS market. And users wouldn't have to run dedicated iPad or iPhone apps to watch their favorite shows—they would watch content right inside their browsers instead.

EVOLUTION OF BROWSER

At the height of the debate over the longevity of native software apps against the power of the web, Mozilla, developer of Firefox browser, claimed that its new browser for smartphones would contribute to the death of mobile app stores. "In the interim period, apps will be very successful," said Jay Sullivan, vice president of Mozilla's mobile division. "Over time, the web will win because it always does." Web proponents such as Mozilla and Google vied for Internet standards that would enable any app to run on any device, just as Java proponents had touted a "write once, run anywhere" mantra back in the 1990s. Ironically, here, they also quoted Adobe's Flash as the software that had emerged as a cross-platform environment for creating animations, games, and apps for the web.

However, for Adobe, creator of a widely-used technology called Flash, which managed video and animations on many websites, fame came through a theatrical feud with Apple when Steve Jobs called Flash software a buggy, battery-sucking relic of the past. Flash was Adobe's highly popular platform for displaying interactive graphics, animations, and multimedia within a web browser. According to Adobe, 98 percent of desktop computers supported Flash, which had subsequently led to its widespread use among application developers. Jobs, on the other hand, called Flash the number one reason for crashing Mac computers. So he didn't want to reduce the reliability on iOS products like the iPhone and the iPad. No Flash, however, meant that the iPhone browser would be incapable of displaying a large portion of the web.

As free Flash was not supported, videos couldn't be streamed from the vastly popular television and movie sites like Hulu, because websites that used Flash to render content or navigation wouldn't work on the iPhone. Adobe's answer: "when you're displaying content, any technology will use more power to display, versus not displaying content. If you used HTML5, for example, to display advertisements, that would use as much or more processing power than what Flash uses." However, many developers complained that the software programs like Java and Flash exhibited bugs, performance problems, and security vulnerabilities, among other issues. Java's promises of universality hadn't quite worked out because different implementations of the Java Virtual Machine meant that Java coders needed to rework their apps for each target device.

Jobs' foresight on the unsuitability of Adobe Flash for the mobile environment and his predictions of the platform's demise came true just a month after his death. In a shocking 180-degree twist, Adobe announced in November 2011 that it would no longer be working to adapt Flash Player for mobile to new browsers, OS versions or different device configurations. Instead, it would focus on building mobile applications based on HTML5—long considered a rival standard. Adobe was now moving to create HTML5 and other web-based tools for desktop and mobile applications. It wasn't just Apple; Google and even Microsoft embraced HTML5 and dubbed it the future of the web.

In retrospect, Adobe had a huge opportunity in taking Flash to mobile devices, the opportunity it bungled and eventually scrapped. The Flash Player plug-in was nearly universal on the desktop and laptop computers and provided a high-performance and cross-platform experience. Nevertheless, HTML5 was not so desperately needed on the desktop web environment. Instead, it was the mobile platforms where Flash was not available and that desperately needed HTML5. Web proponents maintained that the wider acceptance of next-generation Internet standards, particularly HTML5, would win out where Java failed. After all, there were tons of applications that could be delivered through the browser at stunningly low costs. During the web's supremacy that spanned many years, the browser had become a platform of choice mostly due to the economy-of-scale reasons.

With Google, Apple, Microsoft and the rest of the industry lining up behind the HTML5 platform, it seemed certain that the industry had made a clear-cut choice for the near future. Technology companies and application developers alike proclaimed HTML5 was the cornerstone of web's second act. Both Apple and Google, with very different operating ethos, got squarely behind this developer language. Google's affinity for this free-for-all open standard was no-brainer. The HTML5 technology could shift the focus from proprietary software offerings like apps to web-based tools, and web environment was the core of Google's business model: Gmail, Docs, Maps, and so on. In other words, anything that was web-based gave Google the opportunity to do what it did so well. Moreover, HTML5 as an open web standard would enable every browser implement Google product features nearly the exact same way and the development would be quicker and more cost-efficient.

However, it was Apple's decision to not include Flash for the iPhone and the iPad tablet that really set the stage for the rise of HTML5. On the surface, Apple's stance might seem counter-intuitive, considering iTunes, which included the native App Store, continued to thrive with some 160 million registered users in 2010. Research firm Gartner estimated that Apple garnered US$4.5 billion in revenue from apps in 2010. On the other hand, HTML5 promised to pave the way for the dominance of web-based solutions over natively developed apps. So why push for an open web specification that could one day affect the company's bottom line? Apple must have found the answer walking in the corridors of technology history where companies going against the flow of the open standards often ended up as losers.

Moreover, for Apple, preserving user experience was of utmost importance, and in 2010, HTML5 was poised to become a viable platform to solve many of the problems miring the web. Apple seemed so confident in its ability to offer a superior user experience with the iOS platform that having a new web platform that provided a level playing-field and was not run or monopolized by a single company was still a safe bet for it. On the contrary, if Apple couldn't technically lay claim to it, someone else would, and it could well be Google or Microsoft. Also, while the hype behind the quickly-emerging web specification had reached a fevered pitch by 2010, HTML5 and associated technologies still had some growing to do. So until HTML5 caught up with processing and bandwidth challenges, the mobile Internet playing-field would remain more suited for widgets.

In terms of web video, HTML5 technology had already started to take off by the close of 2010; 54 percent of web video became available for playback in HTML5, a significant rise from a meager 10 percent in January 2010. However, though it was getting better in terms of overall performance, the embryonic technology was still miles behind native applications on mobile devices. An iPhone app could outperform a web app doing the same thing by up to 100 times. The biggest difference between HTML5-based mobile web apps and native apps was that the native variety had quick and easy access to a smartphone's hardware features. Native apps generally had an edge in gaining access to platform-specific hardware features such as navigation using a phone's GPS and accelerometer or taking pictures with a phone's camera. Functions as simple as a clock, vibrator, gyroscope, storage, camera, and power management were far more difficult to implement in pure HTML5 apps.

Here, the HTML5 proponents pointed to the fact that, if a particular hardware feature became popular, standards to implement that feature in the browser would quickly follow anyway. They added that it might take some time before HTML5 could handle large data sets and low-latency type situations to be able to do the types of local storage that we saw native apps doing. In fact, there hadn't been much of a debate on HTML5's merits in shaping the web's future; for instance, how HTML5 would enable the browser to rival apps with simpler tasks. However, many watchers in the mobile industry were convinced that it would save the web and render native platform-dependent apps obsolete.

In the early 2010s, they believed that HTML5 was on the cusp of ubiquity and that it would soon become the dominant development stack, taking the mantle from the native apps that had come to dominate the iOS App Store and Android's Google Play. HTML5 was a flag-bearer of the "write once, run everywhere" mantra. Because most browsers functioned in the same way, one app could run on almost all browsers, unlike native apps, which were operating system-specific. Moreover, HTML5 allowed for constant updating without the need for an app store. Every time mobile users logged into the web app, they got the most recent version of the program. The sky was the limit for HTML5, but in all likelihood, it wouldn't be ready for the prime time before 2012, or so they said.

THE HTML5 ANTI-CLIMAX

In October 2011, a Facebook engineer stood in front of Mark Zuckerberg and used a whiteboard to illustrate how Facebook was facing an imminent danger. Cory Ondrejka told the social media firm's young chief that the problem was Facebook's mobile app running on Android and iOS platforms. Facebook had built its mobile app using HTML5 software instead of native iOS or Android code, and over time, it had turned slow and awful to use. Ondejka told his boss that he could rebuild Facebook's app using native code and make it much faster. Nine months later, Facebook released a new version of its mobile app, and the technology press and mobile users were unanimous in their praise for its speed and usability.

Facebook, arguably the biggest champion of HTML 5, had been experiencing slower performance on mobile devices with HTML5 and mobile browsers compared to native apps. Facebook had been the quintessential leader in HTML5 development, and was among the early companies to open up to this innovative software technology. However, after the difficulties that Facebook faced in developing mobile apps, Zuckerberg was forced to call HTML5 "one of the biggest strategic mistakes we made." Facebook with its hacker culture was a beacon for mobile developers, and now with this fiasco, HTML5 had actually taken a step back in developer acceptance.

Another rude awakening came with the iPhone maps fiasco in 2012 when Apple replaced Google Map software with its own HTML5-based web app. Apple made a business decision by going the HTML5 way but could not pull it off on its own. It was another stark reminder of the fact that HTML5 apps were far behind native apps. Still, Facebook and other social media outfits like LinkedIn continued developing HTML5 and other web technologies, and they seemed committed to mobile web in the long haul. It's not that HTML5 was a failure. In retrospect, Zuckerberg's regret was most likely based on the fact that Facebook spent two years dithering on HTML5 when it wasn't simply ready.

A website like Facebook was constantly updating and changing, so an HTML5 app seemed a no-brainer. Instead of having to wait for approval from Google Play or Apple's App Store, the web app would simply update itself. For developers, it eliminated the need to have to rewrite and resubmit the app every time it needed updating. And for mobile users, it freed them from having to reach for the "Update" button every few weeks or months. However, as Facebook engineers found out the hard way, HTML5 was still in the adoption phase and was outpaced by the native frameworks. Still, Facebook wanted to reach everyone in the world, and that could be achieved through both native apps and HTML5-based web apps. Facebook had been instrumental in setting up the Mobile W3C Community Group to promote the development of mobile browsers. However, Apple and Google, who held more than 85 percent of the global smartphone market and had a vested interest in the native app ecosystem, never signed on.

Some industry circles partly blamed Apple for the blowback against HTML5 in 2012. The maker of iPhone and iPad had limited the UIWebView availability in Safari browser, causing hybrid and web apps to perform slowly in comparison with native apps. UIWebView was effectively a miniature web browser that developers could embed within their iPhone and iPad apps. The user could view a web page within the app itself instead of launching the separate Safari app. Apple was also a usual suspect because its App Store was one of the primary reasons for people buying iPhones and iPads. Mobile web apps that circumvented the App Store were apparently detrimental to Apple's bottom line. However, there were a number of stumbling blocks slowing performance of web technologies like HTML5 relative to native software.

For a start, JavaScript, a millstone in web technologies and the de facto scripting language of apps in the browser, was an interpreted language. A JavaScript source code acted like a set of instructions for a piece of software called an interpreter, rather than being compiled into machine code. It was this feature that gave JavaScript portability between platforms, allowing it to run on a browser regardless of whether that browser was running on a Mac, Windows or Linux PC, or an Android or iOS phone. However, JavaScript's additional level of abstraction from the underlying hardware, compared to software languages that compiled to machine code, came at a cost to performance.

Over the time, browsers had started implementing JavaScript engines that slightly lessened this performance gap by using just-in-time compilers, which typically compiled certain stretches of JavaScript into machine code just ahead of being run. However, common tasks that people might want to carry out on a phone or tablet—storing photos and voice recordings locally, making calendar entries or navigating using a phone— were still too difficult to implement in a mobile web app. There was also the problem of the web platform itself not being consistent; the same app might look or behave differently across different browsers. There were differences in what mix of web technologies—HTML, CSS and JavaScript— and features each of the major browsers supported, leaving developers to have to implement workarounds for missing features in different browsers.

Even when the same technologies were implemented across browsers, there were cases of these technologies being implemented differently. Some HTML5 features were supported in Chrome but not in Safari or Firefox. It was imperative for mobile browser makers to ensure that they implemented web technologies in the same way so that web pages and apps would look the same in every major browser without developers having to fix quirks. Moreover, the HTML5 versus native debate mostly focused on operating systems and the challenge of building great apps for a multi-platform world; it masked some of the real problems that the industry was facing when it came to mobile apps. For instance, were there appropriate back-end architecture and business analytics available for mobile web apps? The standards for middleware and back-end data access that defined the web era didn't work that well for mobile.

App developers were apparently swinging away from the web and HTML5, and they were leaning even more toward native apps. That's because the HTML5 apps they built weren't performing. As a result, native apps continued to dominate mobile usage in the mid-2010s, and that threatened the momentum of the development of HTML5-based mobile web apps. Apps were considered a fad in the early 2010s, but they had now completely dominating mobile usage. And the browser was just one more widget in a sea of mobile apps. Apparently, apps had won and the mobile browser had taken a back seat.

According to a 2014 report produced by mobile analytics specialists Flurry, native apps commanded about 86 percent of U.S. mobile users' time, about six times more than the mobile web did. When it came to performance of apps, for instance, games set the benchmark and HTML5 was just not ready for games. Games commanded 32 percent of time spent on mobile devices. And, games could do much more within a native app than HTML5, and so could apps that used the device's camera or other built-in features. Web apps were not a match for native apps in many areas. For example, running web apps offline was a sub-par experience and limited access to smartphone hardware made simple tasks like saving photos more difficult than it should be.

The HTML5 camp was fighting back to make up for the lost ground. For a start, HTML5's application cache (appcache) feature was touted for getting offline to work on web apps smoothly. Although, in the early going, developers found it difficult to deploy it in the manner in which it was intended to be deployed. Local storage was another area being lined up for an overhaul with the publication of a definition for a new web storage API, which would allow megabytes of data to be stored by apps and sites running on the browser.

NATIVE APPS OR HTML5?

In 1994, when Amazon started as a tiny mail order company that sold books on the Internet, the genius move on its part was to remove the middle man from the buying value chain and let the user key in the order into online entry system directly. Before the web, the only way to perform such tasks was through a piece of specialized software that the customer would have to install on his PC. And there were only a few industries, such as banking IT systems, where these specialized software apps were used. That's because the installation of apps on PC was considered time consuming and painful for both the IT departments and the end-users. However, when people shopped at Amazon, they didn't need to download software to do it; they just visited the Amazon website.

When Amazon.com came of age as a digital retailer, web proponents declared apps the walled garden-style silos like the CD-ROMs of the 1990s. However, that sense of victory was rather short lived. A decade later, people who shopped on

Amazon website were using apps like Price Check rather than the mobile web, even though Amazon's mobile web capability was perfectly functional. Why were smartphones and tablets dependent on installable software when people had spent many years trying to get rid of native, installable apps on the desktop? There were a number of factors that contributed to this reversal of fortunes.

First and foremost, as explained earlier in this chapter, it was harder to build good user experiences using the mobile web compared to native apps. The native toolsets provided by Apple and Google were designed to showcase the contents to their maximum advantage. Moreover, when Apple created the App Store and the developer tooling to go with it, it fixed the problem of local installation before it did anything else. So the apps just worked flawlessly. App platforms like Apple's iOS were dominant also because developers went there to make money. Furthermore, apps were dynamically generated views of data that a user could manipulate directly. On the other hand, web browsers were primarily optimized for the display of documents, and apps were not collections of documents.

And that was the primary reason rich native apps provided a significantly better user experience over web apps loaded via a mobile browser. However, within the web domain, it's also important to make a distinction between web apps and websites. Web apps were a piece of software while websites were made up of content; the two served separate purposes. For this reason, mobile apps were not likely to fully displace the web as the world's primary publishing platform. Content would continue to be primarily published via the web. Take the example of Uber, the cab ordering service that allowed city dwellers to quickly and easily get around through an app on their iPhone or Android devices. However, once the service became popular, the San Francisco–based upstart launched a new mobile website m.uber.com to allow Blackberry and Windows Phone owners to use the service and thus expand the potential user base.

Uber was apparently a mobile app-centric company, but its website was also central to how it marketed itself as a lifestyle brand. Moreover, mobile websites were inherently more suitable for specific user groups like stock market professionals. So, the power of apps aside, developing a mobile-friendly website would still be the foundation stone of a number of businesses' mobile game plan. For instance, HTML5 and browser-based websites and services made a lot more sense for some content providers. Not surprisingly, therefore, HTML5 had scored some small victories in the publishing world.

Case in point: HTML5-driven websites of *Financial Times* and *The Boston Globe*. The British newspaper *Financial Times* took its app out of the Apple's App Store and created a mobile website built entirely of HTML5. *The Boston Globe* was another prominent newspaper who embraced the mobile web and HTML5.

BostonGlobe.com was the paper's answer to the mobile revolution, and it was designed to fit on any device through responsive design.

Responsive design saw a boom in 2012. It addressed one of the main problems in developing content for mobile devices: making content look good on a variety of screen sizes. Smartphones tended to have screens that varied from 3.5 inches to 5.3 inches while tablets ranged from 7 inches to 10.1 inches. It was impossible to develop an app that would look great on all of these devices. And building apps for a variety of screen sizes drove developers crazy. The key to responsive design was that it reformatted content to adapt to the screen it was on. Responsive design was built from a web technology stack that included HTML5 and CSS to create websites that responded to a variety of screen sizes by automatically resizing windows to fit a particular screen. The purpose was to build one set of codes for a website that allowed it to work on multiple devices without having to build separate sites for individual devices or mobile operating systems.

There were other favorable signs for the notion that the web would live on and prosper in the mobile era. For instance, HTML5 seemed ahead in weather and shopping apps, both of which relied more on user analytics, which web-based apps could access faster. Moreover, while native apps were clearly transforming the gaming and social media industries, in certain areas, such as public utility services, web apps still offered some good merits. In other words, mobile web apps had been gradually improving. A proof of that progress came in 2013 when Amazon began letting developers charge for HTML5 apps in the Amazon Appstore. The move showed another step toward web apps being viewed as a viable alternative, or at least a complement, to native mobile applications.

A large number of enterprises was now building hybrid HTML5 apps, meaning the bulk of the app was written in HTML5 with a native wrapper to improve performance, add camera access, etc. An app that could be developed in HTML5, then wrapped with native code, and deployed to various app stores was called a hybrid app. In other words, hybrid HTML5 apps could be accessed through mobile browsers like Safari or Chrome and could also be wrapped in a shell of native code and deployed for the native app stores. This hybrid approach—half browser, half native—could give developers and enterprises an easier route to develop both mobile web and native apps.

The web's cross-platform reach made it a viable platform. There were nearly 8 million HTML developers out there, compared to hundreds of thousands for iOS and Android. So, eventually, HTML5 could gain traction because of its three key merits. First, it was fully backed by the industry. Second, mobile players in the emerging markets were betting that HTML5 could anchor the low-cost smartphone economy. Third, with a majority of mobile operating systems sporting HTML5-compatible browsers, the time would eventually be ripe for HTML5 apps.

Industry luminaries like Chris Dixon were concerned that the increasing use of mobile apps would eventually lead to a future in which the web could become a niche product. Moreover, proprietary walled gardens run by a couple of industry giants would threaten the existence of open web. However, on the other side of the argument, it was also true that that Foursquare, Instagram, WhatsApp, and a number of other success stories could never have happened with just the web. The HTML5 versus native apps techno-religious war was destined to go on for a while. The mobile Internet pond was huge and it was plausible that apps and web platforms would eventually complement each other, with the open web filling the gaps left by specialized apps. In other words, the death of the mobile web was greatly exaggerated.

In the end, the conundrum of having either the browser-based web or the native apps could prove to be just a transition phase. Eventually, a new setup could combine the best of apps—run by gatekeeper-style platform owners—and the web boasting the openness and lack of proprietary standards. It was plausible that, over time, the mobile ecosystem would be split between games and utilities that functioned best on native platforms and more traditional content and local stuff that used the mobile web. While processor-heavy apps reliant on device APIs would be better written with native code, content-driven apps could be created more smoothly on the mobile web.

Mobile users were not bothered with the apps versus browser tug-of-war; they were just inclined toward the best content and services. So the HTML5 community had time on its side. However, only way to disrupt the app economy was to improve the capabilities of the mobile browser. The next chapter will provide a detailed treatment of the mobile web browser technology and business, and will analyze mobile web browser providers' efforts to offer a top-notch, app-like experience.

Apple board member Arthur Levinson is known to have convinced Steve Jobs to permit third-party apps on the newly launched iPhone platform. Jobs was initially concerned that outside apps could affect the user experience with quality and security constraints. He eventually opened up to the idea of outsiders writing apps for the iPhone but put in place strict standards and approval criteria. Apple appointed Levinson chairman of the board after Jobs passed away in October 2011.

In April 2012, the HTML5 specification author Ian Hickson quit the W3C effort to concentrate on editing the alternate HTML standard documents maintained by the Web Hypertext Applications Technology Working Group (WHATWG), a splinter organization of the W3C. WHATWG dropped the version number from its HTML spec, and began calling it "a living standard," whereas W3C continued working on a more static version of HTML5 called "snapshot." Critics of WHATWG's approach said that the whole idea of a living standard was not realistic because it would be impossible for browsers to maintain compatibility with the HTML standard if the specifications were in constant flux.

5 THE BROWSER WARS

"If HTML5 really starts to take off, then it certainly is possible that mobile browsers could become much more significant. That world is not here today, but it's one that people are betting on for tomorrow."
— David B. Yoffie, a professor at Harvard Business School and co-author of a book about the browser wars of the late 1990s, in an interview in December 2012

By the mid-2010s, the industry was beginning to crack the hardware puzzle for HTML5 apps, and Mozilla was on the forefront with its Firefox operating system. In 2012, after Facebook ceded the role of HTML5 community leader, open source developers spearheaded by the likes of Mozilla became the leaders driving HTML5 for the foreseeable future. Mozilla was probably the only company working directly on mobile web browser-related problems with its Boot2Gecko smartphone operating system. Mozilla, who had released the Firefox web browser in 2004, announced an HTML5-based mobile operating system for smartphones later in 2012.

The web software firm had carried out a development effort of two years to build an operating system which started as a set of code called "Boot2Gecko" and eventually turned into an open smartphone platform. It was a Linux-based open-source operating system for smartphones and tablet computers. The new Firefox operating system used HTML5, the lingua franca of the web, and was built on open APIs intended to be shared with app developers. Mozilla had created what it called Web APIs, which tied its browser-based mobile operating system to hardware components like camera and power management.

One of the major challenges in bringing the power of the web to smartphones was to emulate the same user experience that "native" operating systems like iOS offered. And that required tying phone's hardware parts like camera and GPS to the web using HTML5, not specific code languages like Objective-C used by Apple's iOS apps. And Mozilla couldn't afford to wait around for hardware to get better. It needed to make the web stack better, so that it could work on even barebones phones, including in areas of limited or no bandwidth. To overcome these hurdles, Mozilla had created the Web APIs that tied the web browser to the phone hardware. Initially, Mozilla offered thirty Web APIs that commanded features like the proximity sensor, phone vibration, push notifications and power management.

Mozilla combined the offline capabilities with device APIs that allowed web apps to access hardware features such as cameras, battery level and microphone. It allowed web apps to be offline from the start; moreover, it enabled the use of equal or less bandwidth than native apps. Consequently, in 2013, just when the pundits had started to talk about the decline of HTML5, the web standard actually began to make some tangible progress. The Web APIs created by Mozilla provided HTML5 much-needed boost to help run apps on smartphones and tablets.

Apple's iOS or Google's Android were mobile operating systems built with "native" code specific to that platform. They featured browsers that handled quite a bit of a smartphone's basic functions, but these browsers still lived in proprietary ecosystems. On the other hand, Mozilla called web the operating system and Firefox OS-based smartphones "open web devices." The Firefox software wasn't just a browser but a full-fledge operating system where anyone who could write code to design and build a website could also write code to design and build an app.

Mozilla's disruptive mobile OS strategy was aimed at enabling extraordinarily cheap smartphones that would entice first-time web users on the Internet. Not surprisingly, therefore, Mozilla planned to target Firefox OS on the emerging markets that were not already saturated with smartphones. Alcatel, Huawei, LG, and ZTE were among the early handset manufacturers to have joined Mozilla's attempts to open up the browser market. Next up, mobile operator Telefonica announced to bring Firefox operating system to Spain. The other mobile carriers who supported Firefox OS included China Unicom, Deutsche Telekom, Hutchison Three Group, Indosat, KDDI, Sprint, Telenor, Telkomsel, and Telecom Italia Group.

The San Jose–based browser maker aimed to disrupt the walled garden-like app stores with innovative new ways of accessing HTML5 web apps. There was no such thing as a "native" app in the Firefox mobile ecosystem. Instead of having to develop specifically for mobile platforms like Android or iOS, the web itself was the platform for Firefox OS software. If an object existed as a web page, it could easily be changed into an app for Firefox OS platform by essentially turning it into a shortcut for the browser-based operating system. Mozilla was now the web's laboratory for the next generation of mobile products. Its Firefox Marketplace was an exclusive web-based app store that showed off the power of open web devices. It showcased, for instance, Nokia's HERE maps that featured offline capabilities, directions and more.

Mozilla's technology chief Brendan Eich saw the smartphone duopoly of Android and iOS akin to the monopoly Internet Explorer held over web browsing in the late 1990s. He was instrumental in breaking up that monopoly. Eich was at the heart of the Open Web movement that Mozilla had championed since the late 1990s when Microsoft's Internet Explorer took over the web browser market. He had played a key role in the creation of Mozilla, the corporation, which was wholly owned by the non-profit Mozilla Foundation. It was also Eich who had invented the JavaScript coding language, which ran much of the functions on the web, and subsequently had helped build the Netscape Navigator browser. Eich had joined Netscape Communications Corp. in April 1995 after a seven-year stint at Silicon Graphics.

Eich had developed JavaScript for the Netscape web browser. At that time, Java software had been around for five years and Netscape was the first Java licensee. Eventually, the Firefox browser was born out of the source code of Netscape Navigator. In 1998, he helped found mozilla.org, and later in July 2003, when AOL Inc. shut down the Netscape browser unit, Eich played a crucial role in the spin-out of the Mozilla Foundation. Two years later, when Mozilla Foundation turned into Mozilla Corp., he became chief technology officer of this commercial entity. Subsequently, tech cheerleaders like Eich put Mozilla on the forefront of the battle of open web versus apps-centric closed ecosystem. He briefly acquired the role of Mozilla CEO in early 2014, but was forced to resign amid a controversy over his funding for an anti-gay organization in California.

ANATOMY OF MOBILE BROWSER

The mobile browser scene of the mid-2010s was becoming somewhat reminiscent of the browser wars of the 1990s when

Internet Explorer and Netscape Navigator fought for dominance on the personal computer. This time, though, the struggle was shaping up to be over which companies would control the mobile Internet goldmine with browsers on smartphones and tablets. Mobile browsers gave these companies more control over how consumers used the mobile devices and data about how consumers used the web, which these companies could use to improve their services. Furthermore, faster browsing led to more web activity, which in turn led to more revenue for these companies, whether it was search on Google or shopping on Amazon.

Take the case of Rockmelt, the free iPhone and iPad app which tried to make browsing faster and more convenient by opening up a full screen filled with squares of content based on mobile users' interests gleaned from their Twitter and Facebook profiles. Rockmelt also attempted to overcome network latency issues by preloading content on its servers and piping it down to users. In 2010, former Netscape engineer Tim Howes and former OpsWare employee Eric Vishria developed this mobile browser with a social twist that was essentially based on the Google Chromium project, the open-source engine that powered Chrome browser. In August 2013, Yahoo! acquired Rockmelt with an intention to incorporate the browser technology in its various products.

Futureful was another poster child of a new generation of mobile browsers that vowed to bring creative features to the small screen in hopes of luring iPhone and Android users away from the default browser. The Finnish startup emphasized browsing without typing by using mobile users' Facebook or Twitter interests and preferences. The augmented reality-centric predictive discovery engine used social media profiles to predict what users wanted to check online, and brought that information by recommending content in little tags, which floated on the top of the screen. Then, there was MoboTap, the company that made the free Dolphin Browser, which stood out with mobile-friendly features like the ability to set up unique gestures. Mobile users could tap a little dolphin icon at the bottom of the browser screen and activate functions such as searching, browsing, opening new tabs, and sharing web pages.

The history of mobile browsing was quite recent just like its carrier, the Internet. In 1994, the Telecooperation Office (TecO)—a research group at the Karlsruhe Institute of Technology in Germany—developed the first mobile browser PocketWeb for Apple's much-hyped Newton handheld. The product was eventually launched as the first commercial mobile web browser NetHopper in August 1996. NetHopper was a text-only web browser that turned raw HTML coded documents into formatted screens of text on Newton handheld computer. Next year, the British firm STNC Ltd developed a microbrowser HitchHiker, which was unique in the sense that it intended to present the entire device user interface and that it ran on a GSM application stack. In 1999, Microsoft acquired STNC and turned HitchHiker into Mobile Explorer 2.0.

A mobile browser in itself was not a lucrative product. The companies like Firefox made money from search engines like Google and Bing that paid when people used the search bar built into the browser. Opera Software, another browser maker, had also set up partnerships with companies for in-browser search and shopping bars. Oslo, Norway–based firm also made money through licensing agreements with companies like Nintendo which offered its mobile browser on Wii consoles. However, the majority of smartphone users preferred the browsers that came built-in with the handset.

In January 2014, according to web analytics firm Net Applications, Apple's Safari browser captured nearly 61 percent of the mobile browser market, while Google's Android browser had more than 21 percent. Opera Software's Opera Mini browser came in third with about 10 percent of the market. Still, by early 2010s, a slew of mobile browsers were trying to break into a market that was dominated by smartphone powerhouses Apple and Google. That's because the exponential growth in the use of smartphones and tablets led these mobile browser upstarts to see an irresistible opportunity to innovate and turn this little marvel of software engineering into a pocket-sized gateway to the web.

Mobile-browser underdogs like Opera and Rockmelt were quick to point out that the market leaders—Apple and Google—had just taken a desktop web browser and plopped it onto a smaller screen, creating an experience that wasn't great for smartphones and didn't take full advantage of mobile devices' touchscreens. So a new generation of startups—like Opera, Rockmelt, Dolphin, and Futureful—was challenging the built-in browsers on the iPhone, iPad, and Android-running smartphones and tablets, hoping to grab a percentage of the growing market for surfing the web on the smaller screens. They were reimagining the browser space with a myriad of innovative new products.

In mid-2010s, browsing the web on a mobile device was still inferior to using the desktop web or smartphone apps. As explained in the previous chapter, mobile apps were faster to load and were better optimized to small screens. However, web technologists were confident that mobile browsers would eventually improve once HTML5 became pervasive. Their confidence apparently stemmed from HTML5 technology's ability to make websites as functional and visually rich as native apps. An HTML5-focused mobile browser, for example, could allow mobile users to do more complex things like shopping or games on a mobile device. The importance of a robust mobile browser was further magnified when it became the interface to the cloud.

Microsoft, one of the major players in the cloud arena, was betting on HTML5. The software giant was a big proponent of HTML5 also because it needed to embrace the web in order to compete with mobile app incumbents Apple and Google. In fact, Microsoft had bought all of Netscape's patents from AOL for US$1 billion. That included crucial web patents such as cookies, JavaScript, and Secure Socket Layer (SSL). Next, the Redmond, Washington–based IT behemoth began encouraging software developers to build websites that were more app-like. Microsoft's new browser allowed readers of news articles to swipe to turn the page instead of touching "next page" with a fingertip. Internet Explorer 10 had come a long way in its bid to adopt HTML5 technology. And Microsoft continued to publish new improvements for Internet Explorer 10 that made it an even more capable HTML5 browser.

PROFILE OF A BROWSER PIONEER

The opening section of this chapter has profiled Mozilla as a stalwart of the post-iPhone era mobile web business. Another poster child of the mobile web revolution was a tiny browser company based in Oslo, Norway. This Norwegian outfit had practically invented cloud compression to improve mobile user experience through its flagship product: Opera Mini. The compression-happy Opera Mini browser came during the pre-iPhone days and quickly became the best way to surf the web on the Internet-capable phones. Opera Mini requested web pages through company servers, which processed and compressed them before sending them to the mobile phone, thus speeding up transfer by two to three times and dramatically reducing the amount of data transferred.

In 1992, Jon Stephenson von Tetzchner and Geir Ivarsøy were working for Norwegian Telecom Research, the research arm of the Norwegian state phone company Televerket, which later became Telenor. The research group had been developing ODA, a standards-based system for storage and retrieval of documents, images and other content. In 1993, when this team of software engineers established the first Norwegian Internet server and homepage, it felt that the Mosaic browser had a too flat structure for it to be used effectively in browsing the web. So the group put the ODA project to work and started building new document browser from scratch. The first prototype was up and running by late 1993, but Televerket faced a dilemma: the browser program didn't fit in its core business. Next up, in 1994, Televerket became a stock-owned company and was renamed as Telenor next year.

Eventually, Tetzchner and Ivarsøy were allowed to continue development of the browser on their own while using the offices of Telenor. In 1995, they obtained the rights to the software and founded Opera Software ASA while still based in Telenor's office building. Their product was initially known as MultiTorg Opera and was quickly recognized by the Internet community for its multiple-document interface and its sidebar which made browsing of several pages at once much easier. In 2006, the company launched Opera Mini browser that used server-side compression to squeeze data by up to 90 percent, which meant faster page loads and less data usage. It literally brought the web to almost every Java-enabled mobile phone.

Meanwhile, the mobile web landscape had been evolving at a relentless pace and the Norwegian browser company was not sitting on its laurels. In summer 2012, it launched the pay-as-you-go web access service it called Web Pass. Opera was aiming the sponsored Web Pass service at mobile phone operators as a channel for smart pricing. The idea was that mobile users could buy prepaid packages for data-intensive apps like Spotify and Pandora, and these data "top-offs" would carry out file-compression services to make the most of the prepaid data packages. Opera facilitated Web Pass service through a simple click on the Opera Mini browser that routed network traffic through virtual servers, which in turn, compressed the size of files to run on slower networks.

The Malaysian company DiGi Telecommunication was the first mobile carrier to start using Web Pass in a bid to convince the subscribers hooked onto its second-generation cellular network to activate data services. Then, in early 2013, Opera acquired its Silicon Valley competitor Skyfire for US$155 million to ensure that the available network bandwidth was there once a mobile user requested a data-intensive service. If there was no available bandwidth, Skyfire would either compress the video before delivering the content or would give a warning to the user before he tried to play it. Skyfire specialized in video optimization and monetization technologies.

Both the synergy and the objective of this merger apparently made sense. Opera, which boasted extensive relationships with mobile operators, clearly aimed to facilitate them through this deal. Operators were making a crucial shift toward software-defined networking (SDN), which in itself was intended to give operators the ability to fine-tune parts of their networks in ways that were not previously possible. Opera's browsers were known for using server-side compression as a way of saving on data costs, while Skyfire used server-side rendering for video, making it a go-to browser for data-heavy applications. Another critical dimension of the union of Opera and Skyfire related to the emerging markets. Opera's sponsored Web Pass service was the key factor behind the acquisition of Skyfire, and Web Pass was primarily an emerging market-focused service aimed at making mobile Internet packages more straightforward for first-time phone owners and inexperienced web users.

Opera's business figures provided some clues. In 2013, Opera claimed that 32 percent of its users were on smartphones, and that took into account the 21.5 million users of Opera Mobile, a product designed specifically for smartphones. Opera Mobile was an Internet browser in the true sense because it made a direct connection with the web page and showed users live content like video with greater interaction. Opera Mini, on the other hand, was a proxy browser which allowed a much smaller amount of data to load from the web page to speed up things. The web content first passed through the Opera servers located in various locations around the world and the servers striped out overheads, compressed content and pre-formatted the web page. Afterwards, an image of the web page was sent to the Opera Mini browser running on an ordinary mobile phone connection.

In a nutshell, the difference between the two browser products was that Opera Mini opted for the company's advanced server compression technology, essentially doing some of the heavy lifting on Opera's own servers before sending the results to a mobile device. Although the quality of the web page might suffer a little bit as a result, it meant that a web page loaded much faster on the move and required less data, an important advantage for users in developing markets. It was startling to see that Opera Mini users outnumbered Opera Mobile smartphone users by roughly 10:1 and that percentage spelled an enormous adoption of the lightweight browser on feature phones. Therefore, while the early adopters of Opera Mini were in Europe and the United States, the Norwegian company saw a huge growth in mobile Internet usage in the emerging markets.

INTERNET FOR THE NEXT FIVE BILLION

In 2013, according Mary Meeker, a partner with Kleiner Perkins Caufield & Byers, there were 1.5 billion smartphone users in the world. However, that number paled in comparison to nearly 5 billion mobile phone users across the globe. All of these mobile phone users were potential smartphone customers. It was this "smartphone for the next 5 billion" premise that made mobile web pioneers like Mozilla and Opera hopeful about the future. Since 2000, the number of mobile phones in the developing world had grown by 1,700 percent, and many of these mobile users were now upgrading to smartphones with data plans that cost as little as US$2. The price of an Internet-capable smartphone had fallen below US$50, and in India, it was possible to get a tablet like the Aakash 2 for US$25.

The explosion of mobile adoption in large markets like China and India had created openings for unconventional mobile players like Mozilla. China was consuming smartphones faster than the United States during the mid-2010s. At the Mobile World Congress in 2014, Mozilla had rocked the wireless industry by announcing the Firefox OS-based US$25 smartphones. Mozilla had made it clear that it was going after first-time smartphone users in the developing economies. According to the ITU, about 16 percent of Africa's one billion people used the Internet in 2014, which was well behind Asia, with 32 percent, and Arab states, with 38 percent. But Africa was the fastest-growing region for accessing the Internet by mobile phone.

The web was a scarce resource in the developing countries, and mobile with its economy-of-scale benefit offered the most viable way to provide Internet connectivity to the world's most under-served markets. The mobile platform had the power to reach the five billion people on the planet who had no access to the Internet, and that pervasive Internet connectivity could improve public health, transportation, business and democracy across the globe. Cheryl Langdon-Orr, a participant of the Internet Society and ICANN, called this phenomenon the "Internet for Everyone."

According to market research firm StatCounter, users accessing the web through mobile devices had been doubling every year since 2009. That further reinforced the notion that the use of the mobile web was permeating the everyday life of people around the world. "The thing to look for in [2014] is that you have one to two billion Android handsets coming on-line," said Marc Andreessen, the man widely credited for the invention of the first popular web browser Netscape Navigator. "We've never had the ability in our industry to reach five billion people with a computer and now we have the ability to do that. That's big." Andreessen, now a Silicon Valley investor, had a stake in the mobile web upstart Rockmelt covered earlier in this chapter.

Take Nokia as a case study of the wireless industry's bid to serve the so-called next five billion mobile web users. The Finnish mobile manufacturer had been relying on this specific industry premise to reinvigorate its smartphone strategy which had been marginalized by the duopoly of Apple and Android. During the mid-2010s, new smartphone connections were mostly destined to come not from North America or Europe, but from Africa, South America, and large Asian markets like China, India and Pakistan. Nokia had been highly efficient manufacturing and logistics operation capable of churning out a dozen handsets a second and selling them all over the world. Nokia also boasted applications and services broadly available for specific regions and languages, particularly in Europe and emerging markets in Africa and Asia.

Nokia's Symbian platform, despite its cumbersome software legacy and steep market share loss, had remained a leader in mobile web browsing till the end, according to a 2012 survey from StatCounter. The Nokia Xpress browser incorporated the firm's cloud technology to reduce data consumption, speed load times, and shrink data downloads by up to 90 percent by compressing web pages before delivering them to mobile users. It was a crucial feature for mobile users who frequently used an older second-generation wireless connection or had a mobile phone contract with a low data allowance. The browser also supported more than 10,000 web apps—like Nokia Nearby—to give browser feature a greater functionality.

Nokia and its former platform partner Microsoft were clearly lagging behind Apple and Google in the native app arena, but the "apps versus web" battle for smartphone riches provided Nokia and Microsoft with a window of opportunity. In September 2013, Microsoft had announced to acquire Nokia's mobile handset business, but as mentioned earlier in this chapter, Microsoft being an HTML5 cheerleader, would most likely carry on the path set by Nokia to leverage the power of the web to lure the billions of new smartphone users.

Nokia Nearby web app tapped the firm's mapping assets and cell-tower positioning technology to offer location-based services to phones that lacked GPS functionality. The web apps like Nokia Nearby could provide the Finnish mobile firm with a level playing field; Nokia could use these web apps to facilitate moderately priced, Internet-ready models in emerging markets around the world. Nokia was also forming a partnership with Mozilla to integrate its location features in the upcoming Firefox operating system.

Nokia's sales figures of the fourth quarter of 2012 clearly suggested that its Asha smartphone line was gaining good traction in emerging markets. The old Symbian devices used to sell quite well in India, Eastern Europe, Africa and Latin America and these were the markets where Nokia likely had a strong opportunity with Asha phones. Nokia's Asha family of low-end, multitouch handsets was barely classified as a smartphone. But calling these devices feature phones also seemed to be a misnomer. The Asha phones could send e-mails, link instantly to Facebook, and download free and paid games and apps from the Nokia Store. Even the camera integrated into the Asha handset was able to automatically resize pictures for easy sharing and posting.

Google had also acknowledged the strategic importance of markets like India and countered by opening up brick-and-mortar retail stores in 2013. The Android Nation stores in India would serve as an experience center for Android devices just like Apple Stores. Next up, in China, local handset makers like Huawei and ZTE were effectively leveraging free Android software and chips from Taiwan–based electronic components supplier MediaTek to produce inexpensive mobile phones. And these Chinese telecom manufacturers had started using their domestic success to expand globally. In 2008, MediaTek supplied complete reference designs for phone chipsets, which enabled manufacturers in mainland China to produce phones at an unbelievable pace. By some accounts, this ecosystem produced more than one third of the phones sold globally in 2012.

INTERNET IN THE CLOUD

In the final part of this chapter about the future of the mobile web, it'd be worthwhile to mention that the promise of the mobile web was intrinsically tied to another rapidly evolving phenomenon: cloud computing. Cloud was a metaphor for the Internet and cloud computing was a phrase that was being used to describe the act of storing, accessing and sharing data, applications and computing power in cyberspace. In the mobile context, the bottom line was that the phone itself could not do it all; it's the power of the phone's processor plus a combination of network and cloud computing that could make it happen. With a giant server farm, the data got passed from smart devices to wireless networks to the cluster, where it got processed and sent back to the device.

In the early 2010s, corporations were starting to experiment with the combination of the smartphone and cloud computing—an innovation that many saw as the most significant since the advent of the Internet. It allowed, for example, a travel agent to facilitate booking and ticket arrangements, provide alerts about canceled flight, and offer plans for alternate arrangements via a smartphone tied to the cloud. Users could access their bank accounts with real-time transactions, payments, and transfers with a single click using mobile phones. The biggest and most popular application could be games with no need to install it on the phone; the processing could be performed by clouds at a higher speed, enabling a breathtaking gaming experience without any glitches or interruptions or processing power limitations.

The number of mobile users that the wireless industry had the power to reach was far more than the number of smartphone users in the early 2010s. The fact that feature phones themselves were becoming more capable with smarter built-in web browsers could have an impact on the mobile cloud's adoption. It could be one way the mobile cloud grew on the cheap. That was also a reason browser was being seized up as an OS alternative of some kind. Eventually, as more apps became truly cloud driven, even feature phones could become more and more cloud compatible. It was this cloud-centric premise that could provide a logical path for all mobile phones to gradually morph into smartphones. Whether it was a feature phone or smartphone might not matter much to users as long as it nicely fit in the category of a cloud phone.

Next, cloud computing promised to address a major conundrum facing the mobile Internet: the net effect. The rise in mobile Internet traffic subsequently translated into more data hungry users and networks transitioning to 3G and 4G systems to accommodate this tsunami of data. However, AT&T's

iPhone data snag was a testimony that 3G—even 4G network—wouldn't be able to meet the insatiable demand that smartphones and mobile Internet could create. Here, cloud came up with a viable solution to take the bulk of the network processing power. The best these 3G and 4G networks could do was to provide a safe and efficient link to the cloud, and cloud would efficiently manage the data usage challenges.

The lack of speedy mobile Internet access would inevitably create barriers in the shift toward mobile computing. While much of the content in the preceding chapter about HTML5 software and its ability to deliver compelling mobile apps focused on the device performance, a potentially bigger problem resided on the network side. A slow mobile network meant that browsing through websites on a mobile device could be like slogging through a marsh. However, the HTML5 made use of local caching in the mobile cloud environment and enabled apps get past network-centric issues.

There was another reason cloud computing could become a disruptive force in the mobile Internet world and that had to do with how apps were distributed. Mobile applications were still indirectly tied to a wireless carrier. If someone wanted an iPhone app, for example, he or she had to first have a relationship with the mobile operator carrying the iPhone. If that person wanted a BlackBerry app, the same rule applied. But with a mobile cloud computing link, as long as the user had access to the Internet, he or she had access to the mobile applications. Moreover, once mobile applications began to store data in the cloud as opposed to on a mobile device, the applications would become far more powerful as their processing power was also offloaded to the cloud. All users would need was a basic mobile Internet connection, and the data could be passed back and forth in near real-time, without having to bog mobile devices with heavy applications and software.

Cloud computing was a hot button among IT circles in 2012, and during the next year, it got hooked up to the notion of big data, which in turn was intrinsically linked to the next big thing in mobile computing: The Internet of Things. It was a major departure in the history of the Internet as connections moved beyond computing devices and began to power billions of everyday devices—from parking meters to home thermostats. Nobody was even talking about the mobile Internet in 2014. Now everyone was talking about the Internet of Things which was the Internet beyond PCs, tablets and smartphones: the Internet of devices that had embedded technology to sense either their internal states or the external environment.

The first phase of the smartphone revolution was about turning cell phones into powerful pocket PCs and compelling mobile Internet devices—like the iPhone made this possible by overcoming form factor and data connectivity challenges. The next phase of this mobile revolution was about taking this mobile Internet juggernaut and spread it everywhere: homes, cars, utilities, television, commerce, and so on. Smartphones still had big gaps in battery life, network connectivity and form factor.

Connected wearables a.k.a. ubiquitous computing would begin to address these issues, first in the form of smart accessories: glasses, bracelets, watches, earpieces, and more. Computers embedded into things like car and refrigerator would be next. Smartphones and tablets displaced the once-dominant PC, and now wearable devices were poised to share the limelight with smartphones.

It was the power of the cloud that made the sauce for this next big thing in mobile computing: wearable devices. Apparently, wearable devices had won a lot of attention from IT companies looking for growth beyond the smartphone market. However, smartphone wasn't exactly a left-over; it would become more of a digital hub for wearable devices. Smartphone makers had to master the tricky art of providing dependable mobile Internet service and now wearable manufacturers could simply piggyback on those innovations using simple Bluetooth or other protocols to communicate with a smartphone and thus with the outside world.

Take Google Glass as an example. In a way, Google had augmented the smartphone and its alter ego mobile Internet to its wearable headwear through the cloud. Many industry watchers called this voice-activated computer-monitor combo worn on eyeglass frames the next iPhone. The devices like Google Glass epitomized the next mobile computing paradigm: connected wearables. The next chapter takes a peek at how the notion of the Internet of Things was converging and colliding with futuristic wearable products like Google Glass.

JavaScript was born out of a need to make web pages dynamic. Netscape wanted a lightweight programming language for its web browser that would complement Java by appealing non-professional programmers. When Netscape hired Brendan Eich in April 1995, he was told that he had ten days to develop a working prototype of a programming language that would run on Netscape's browser.
Image: Mozilla

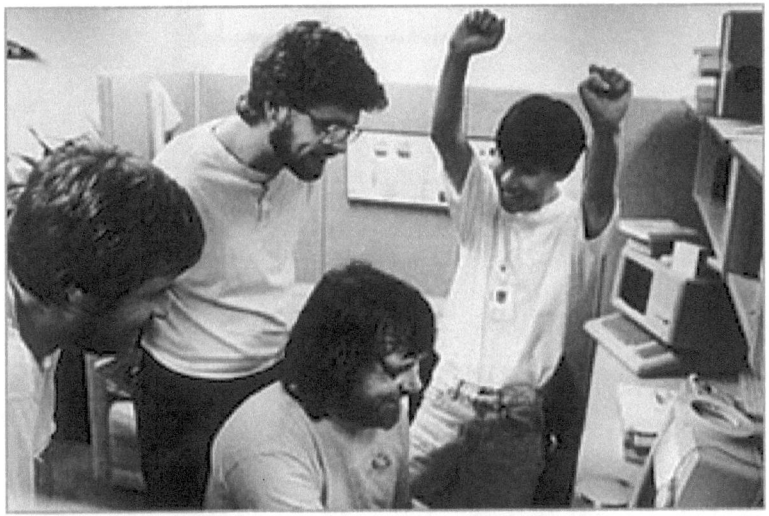

Apple's Newton MessagePad became home to the world's first mobile web browser—PocketWeb—developed by TecO in 1994. A team of Newton engineers is seen in this picture taken in 1992. Image courtesy of Michigan State University

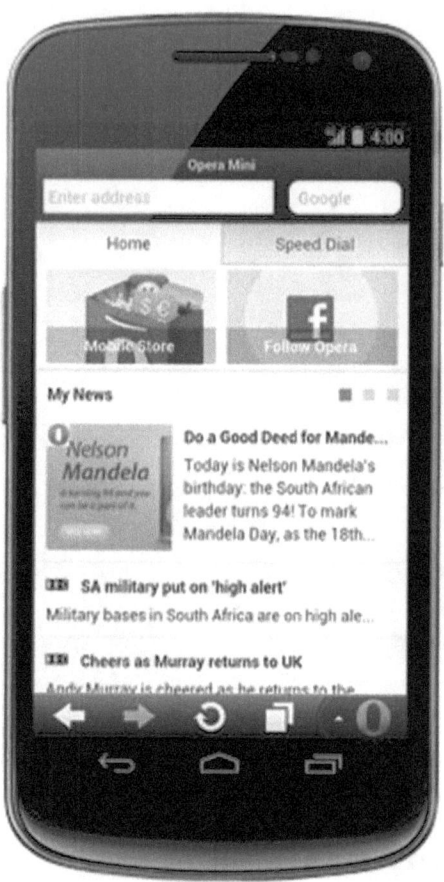

The Opera Mini browser was a popular choice among mobile users because it did some of the heavy lifting on the company servers to compress content before sending the results to a mobile phone. Image: Opera Software ASA

Jon Stephenson von Tetzchner (left) and Geir Ivarsøy founded Opera Software in 1995. The browser work was originally intended as a research project at the Norwegian telecom company Telenor. Tetzchner and Ivarsøy got the rights to the software, formed Opera as a separate company, and began developing browser software for personal computers and mobile phones.

6 DIGITAL SIXTH SENSE

"Connecting products to the web will be the twenty-first century electrification."
— Matt Webb, CEO of Berg Cloud

The wearer said, "O.K., Glass," and the glasses leap into action, performing most of the functions of a smartphone: check e-mail, upload photos to social media, and take videos of the world from the viewpoint of the user's eyes. Welcome to the smartphone 2.0. The futuristic Google Glass could take photos and videos, get turn-by-turn directions, send and receive texts, make and receive phone calls, do Google searches, and share content via Google's social network Google+. The Mountain View, California–based company would eventually integrate Google Now—a virtual assistant that interacted with Gmail, Google Calendar and current location to push information and reminders—letting users know, for example, an hour before a lunch meeting how long the drive to the restaurant would be given the current traffic.

Google Now was the company's attempt to divine and deliver needed information based on context. The voice recognition app gave directions and traffic updates when users asked for them, scanned their calendar, and displayed alerts about meetings and other commitments. Google Now, originally developed for the iPhone and Android phones, made a perfect fit for wearable devices like Glass because it provided instant, even predictive information to mobile users.

Google Glass could connect to any Wi-Fi network or tether to a smartphone's mobile connection via Bluetooth. The newish Internet eyewear also employed a variety of wireless connections—from RFID to infrared to QR codes—to identify a connected device that could be manipulated and used for a myriad of communication purposes. The voice-activated computer-monitor combo worn on eyeglass frames ran on a modified version of Android operating system. However, unlike traditional Android devices, Google Glass user interface was made up of cards that a user could select and interact with, primarily by scrolling through them to the left or right.

Nearly a decade after the Japanese mobile carrier NTT DoCoMo called its i-mode phone a remote control for the world, Google took most of the smarts of a mobile phone and built a new kind of consumer electronics device. In fact, Google took the major components running on an Android phone and glued them into an LCD hooked onto the front of one eye on the pair of safety goggles. Google Glass had packed a smartphone's basic communication, search, and navigation functions into a miniaturized display unit and reengineered itself into a computer-on-the-face. In short, this Internet-hooked futuristic appliance was distinguished by its hyper-integration of an optical head-mounted display, augmented reality, camera, web access, and voice-based interaction.

Computing pioneer Vannevar Bush had presented the idea of wearable computers and cameras in a *LIFE* magazine reprint of his article "As We May Think" published in *The Atlantic* a couple of months earlier in July 1945. However, the history of wearable computing was littered with failed projects since its actual start during the 1980s. So nobody seemed interested in making a wearable computer for the mass market. Eventually, Google realized it would have to make one itself. Google engineers were convinced that wearable devices would be the next big shift in mobile computing. They could see that smartphone, despite all the wonderful things it enabled, took away people's attention and that they couldn't do anything else during its use.

In 2010, the company's top-secret projects lab Google X began the development of this camera- and Internet-equipped wearable computer. The project was announced on Google+ by Babak Parviz, an electrical engineer who specialized on the interface between biology and technology, and had worked on putting displays into contact lenses. Steve Lee, a veteran product manager who specialized in location and mapping technologies, was also involved in the project's initial development. Lee had earlier worked on Latitude, a Google app that enabled users to broadcast their GPS location to friends. Thad Starner, a Georgia Tech professor who had been building and wearing head-mounted computers since the early1990s, eventually became the technical lead for Project Glass. Back in the 2003, Starner had shown Google founders Larry Page and Sergey Brin a clunky version of a wearable computer that he had built at Georgia Tech.

Glass was a pet project of Google's co-founder Sergey Brin. The Internet-connected eyewear was released for developers in February 2013 and would be available for consumers later in 2014. Google Glass was in fact a do-everything computer and information portal that boasted augmented reality technology and epitomized the next wave of disruption in mobile computing. It represented a new class of wearable and embedded computers that first absorbed the smartphone capabilities and then promised to offer even more. Many industry watchers called Glass the next iPhone. It was a great idea that encouraged people to imagine and to create innovative new applications and spawn the brand new wearable industry. The technology behind the Glass could change the world.

On the other side of the fence, however, some critics pointed out that Glass was a product ahead of its time. It was a mini-computer on your face with a social twist; consumers at large were wary of it being a somewhat creepy device that secretly searched information for its owners. Moreover, the product design of Glass didn't go well in the fashion-conscious consumer electronics world where it was imperative for a personal device to look cool. The US$1,500 per pair price tag of Glass didn't help either when it went on sale for just one day on April 15, 2014. Here, unlike the iPhone, which became a consumer icon, Google could first position Glass for professional use cases. The device, for instance, could bolster operations in a truck, train, taxi and boat by providing timely and concise data to help navigate quicker and safer.

Google could take Glass to specific markets and create powerful applications in collaboration with third-party developers. A Google Glass app, for instance, could allow a security guard to unlock doors remotely at the touch of his spectacles, linking to a USB Bluetooth module used in the lock controller. The sweet spots in the business arena could in turn intrigue the public at large about the potential of Glass as a powerful information tool. Nevertheless, it clearly marked a new phase in mobile computing beyond smartphones and attempted to make wearable computing mainstream. And, Glass wasn't the only disruption in the town.

THE NEXT BIG THING

The competitive drama moved to the next episode when industry luminaries like Qualcomm and Samsung and upstarts such as Pebble began vying for a computer-on-your-wrist a.k.a.

smartwatch. According to *Fortune*, Pebble had sold 400,000 smartwatches in 2013 and was on track to double revenues in 2014. In early 2014, Google took the lead in this nascent segment once more by unveiling Android Wear, a new version of the mobile operating system designed specifically for on-the-body devices like smartwatches. Soon after the launch of Android Wear platform, large electronics OEMs such as LG, Motorola and Samsung announced smartwatch projects.

A smartwatch promised the wearers avoid a trip to their pocket to look at their smartphone and represented the first baby step toward a wearable future. Just like Google Glass, the smartwatch chick was another manifestation of Internet-hooked computers that gave mobile users access to instant information. Another common trait that smartwatches shared with Google Glass was that these connected wearables embodied the notion of lifestyle computing and carried the promise to make these tiny computers far more pervasive than smartphones and tablets. The next similarity with the Glass came with the use of microphones and voice control technology that could turn a watch into a mini Siri-like device.

Google's Android Wear, which first appeared on watches, would eventually become the company's platform for all types of wearable devices. Its user interface was based on two core functions: suggest and demand. The "suggest" part of Android Wear used notification cards to make up the watch's context stream. That included text messages that would buzz wrist when they came in and morsels of information like scores of sports games that got silently added to the stack. Notifications could be supplemented with additional pages for relevant information like meeting reminders that people could access by swiping sideways on their smartwatch screen. The "demand" part encompassed a list of commands that could be spoken or tapped on the screen. That included features such as calling cabs, taking notes, sending messages, setting alarms, and more.

By the early 2010s, smartphones and tablets weren't turning heads anymore; their existence was a given. It was now the peripheral devices like Google Glass and smartwatch that were making waves by leveraging the smartphone connectivity. The Internet-connected smart wearables continued to proliferate in the home and beyond, and it was likely that they would create, or disrupt, multibillion-dollar markets in much the same way Apple did with smartphones and tablets. So, mobile Internet stalwarts like Apple and Google were now betting that cars, computers, home devices, mobile phones and refrigerators would communicate with each other, generating insights that could be converted into heaps of data, and subsequently into concise and actionable information.

The connected wearable devices like smartwatches and Glass represented the next phase of mobile computing evolution after the advent of smartphones and tablets. They were an extension of the smartphone and most of these devices didn't aim to replace smartphones. Instead, they worked as satellite devices that amassed useful data or relayed notifications from a primary mobile device. If they had screens, they could display simplified versions of mobile apps. Take Google Glass as an example; it brought the power of the smartphone out of mobile users' pocket and into their field of vision, accessible any time they glanced its way.

Wearable computers were basically small and compact electronic devices designed to be worn by a user. They were also referred to as body-borne computers, considered as a type of wearable system that had in its core an electronic device that performed calculations and processed information. Wearable appliances incorporated more sophisticated applications like retina scans with built-in cameras, use of cameras to measure blood glucose, sensors that provided heartbeat tracking via Bluetooth, and so on. Industry watchers envisaged these wearable systems to become the primary means of accessing the web by the end of 2010s.

The advancements in technologies like augmented reality and voice recognition were making wearable computers less unwieldy and awkward. More efficient and smaller high-capacity batteries now provided enough power to run wearable devices. Next up, thinner and sharper displays made it possible to create more comfortable means through which users could interact with wearable devices. Touch input systems had also become more responsive in the post-iPhone arena. Moreover, processors had become faster without overheating or requiring active cooling systems. Over the years, sensors had also become smaller and more powerful. Furthermore, the chips and sensors were cheaper than ever. In retrospect, the advancements in system-on-chips (SoCs) used in smartphones and tablets allowed massive miniaturization and cost reduction, and wearable computing had taken it from where the smartphone had left.

It was an interesting crossroads: connected wearables were converging and colliding with the Internet of Things and that resonated in what Cisco's John Chambers had famously slated during his 2014 CES keynote as the "Internet of Everything." That pivotal moment in technology history had brought us an entirely new era of products where a new ecosystem of wearable devices was reshaping the Internet of Things into a far more useful phenomenon: the Internet of Everything. Wearable devices were becoming the consumer face of this wider shift otherwise known to the world as the Internet of Things.

Now, there were two versions of the "Internet of Things." The first one was based on sensors embedded into the existing stuff such as a mattress equipped with a sensor. However, for the same task, a user could have a sleep-monitoring wristband, which came under the wearable side of the Internet of Things. In this case, the wristband was hooked onto a server in the cloud over a link that was practically a new manifestation of the mobile Internet. The coming sections will dig deeper into this new technology crossroads.

WEARABLES ARE A THING

The Internet of Things was in many ways the mobile Internet on steroids, the one connected to an incredible variety of handheld, household and industrial objects, each of which could transmit data or even control other connected devices without any human intervention. It was embedded with a massive amount of processing power and connectivity scattered across all sorts of consumer, household and industrial devices otherwise known as the "things" and had become synonymous with the digital sixth sense. These Internet-linked objects could be everything from healthcare monitors to traffic lights to thermostats to trains. They encompassed automobile and factory automation sensors, industrial robotics, sensor motes for increased agricultural yield, and infrastructure monitoring systems for areas like road and rail transport, water distribution and electrical transmission.

The devices on the Internet of Things broadly fell into two large categories: sensors and controllers. In other words, sensors and controllers were the "thing" in the evolving Internet of Things landscape. Sensors monitored anything that could be measured: temperature, location, power, radiation, atmospheric pressure, and so on. These sophisticated devices linked to the Internet of Things were akin to a central nervous system that could serve as the digital sixth sense of the IT world.

The low-cost sensors were a technology marvel that redefined how people interact with the world by producing an unprecedented amount of information in an automated fashion. They resided inside a myriad of electronic devices—cars, traffic light cameras, parking meters, etc.—and laid the foundation of the big data movement. As mentioned in chapter 1, Apple had thrust gyrometers, accelerometers and other sensors through the iPhone, and from thereon, sensors had become synonymous with the Internet of Things because of their sheer volume and pervasive use. The idea of ubiquitous sensors communicating information with humans and among themselves had long been the staple of science fiction stories. Their time had finally come with the advent of the Internet of Things.

In turn, the data that these low-power sensors produced fed their counterpart, the controllers or the objects that responded to the world around them—such as actuators, switches, servos, valves, turbines and ignition systems. There were apparently fewer controllers connected to the Internet than sensors, but they played a crucial role nonetheless. A large part of these controllers was now increasingly made up of wearable devices such as smartwatches, connected fitness bands, and smart eyewear. Connected wearables were becoming the most prominent part of the Internet of Things landscape, and they were taking the pervasiveness of the Internet one step further.

A multi-time zone travel watch, for instance, became a much simpler proposition with an Internet connection. So was the case of a wristband that tracked user's heartbeat and loaded beats-per-minute to the cloud, where an app might help the user store and analyze real-time health data. Wearable fitness bands like the Jawbone UP and FitBit already tracked the activity level, sleep quality, and how many steps a user took during the day. Another case study: LG Tromm washing machine used a smartphone to sense the type of clothes loaded onto the machine and then automatically set the most appropriate cycle. But the device that really made waves other than Google Glass during the mid-2010s related to a Google acquisition: Nest thermostat.

Thermostat was the new kid on the Internet of Things block that used data connectivity and smart software to enable new types of home services. The Internet-connected thermostat employed the home Wi-Fi to learn owners' habits and configured the house's cooling and heating patterns more intelligently. That led to lower utility bills and a more energy efficient home. Nest thermostat wasn't the first smart energy management product to hit the market, but it was a very sleek device and made it simple for consumers to understand and use. Tony Fadell, the General Magic and Apple alumni and the man behind two of the most iconic gadgets in technology history, the iPod and the iPhone, was able to produce another echo of the future mobile computing business through the realization of Nest thermostat. Nest had delivered another crucial building block of smart home when it unveiled the Protect smoke alarm and carbon monoxide detector that replaced the annoying screech with a human voice.

Back in 2005, a UN report had summed up the nascent Internet of Things industry with an uncharacteristic precision:

"The next logical step in this technological revolution (connecting people anytime, anywhere) is to connect inanimate objects to a communication network. This is the vision underlying the Internet of things. The use of electronic tags (e.g. RFID) and sensors will serve to extend the communication and monitoring potential of the network of networks, as will the introduction of computing power in everyday items such as razors, shoes and packaging. Advances in nanotechnology (i.e. manipulation of matter at the molecular level) will serve to further accelerate these developments."

The idea of wearable gadgets and connected objects wasn't new. However, the early wearable computer configurations were extremely bulky and obviously very expensive for the time, so their producers eventually ended up filing for Chapter 11. Even computer industry giants like Apple and Microsoft had been part of this marketing field of dreams that remained in a transitional phase for quite a long time. Microsoft, for example, had unveiled a concept smartwatch at the 2003 CES in Las Vegas based on a technology it called smart personal objects technology (SPOT). The device was sold for a few years by several watch makers without much success. But then the iPhone's touchscreen, swipe controls, and robust mobile Internet connection broke a number of technology barriers. By early 2010s, pundits began saying that wearable computing could replace the smartphone over the next decade.

MANY FACES OF MOBILE INTERNET

Now that mobile Internet had reached a relatively mature phase and the Internet of Things was becoming a commercial reality, what was beyond mobile Internet and even the Internet of Things? A lot! Take, for instance, the Internet of photos. Photos were the most popular piece of personal content that mobile users shared on the Internet at a time when the amount of personal data, videos and even sound files being uploaded to the web was growing exponentially. In 2013, according to Mary Meeker, one of the leading Internet analysts, more than 500 million photos were shared each day on average. And she expected that figure to double year-over-year. The rise of the visual web had turned the upstart Snapchat into a media darling. Facebook and its photo-sharing buy Instagram followed closely behind.

Then, there was this Internet of messaging in the form of popular apps like Viber and WhatsApp. For years, mobile phone operators across the world had made fortunes by charging for a sheer volume of text messages. Short messaging service or SMS, a technology that routed the text through the same infrastructure that mobile operators used to handle voice calls, had long been the main way to send a message from one mobile phone to another. But then mobile Internet applications like

WhatsApp began to offer users a cheap or free way to send messages. Once people switched to smartphones with better Internet access, they started relying on instant-messaging services like WhatsApp to communicate. In 2013, according to market research firm Ovum, mobile operators had lost nearly US$32 billion in revenue when users shifted from text messages to cheap Internet messaging services.

WhatsApp generally charged US$1 a year, and the first year was free. It explored the openness of the Internet, which gave it a leg up over proprietary messaging systems like BlackBerry's BBM and Apple's iMessage, which only worked on their respective platforms. In February 2014, Facebook bought WhatsApp for a whopping US$19 billion in cash and stock in one of the largest deals in the history of mobile Internet. Between Facebook, the world's biggest social network, and WhatsApp, the most popular smartphone messaging service, about 1 billion photos and 30 billion messages were sent per day. And just as with photo-sharing mobile site Instagram, Facebook would continue to operate WhatsApp as a standalone service under the existing WhatsApp brand. Facebook—now increasingly looking like General Electric of the Internet—could eventually turn WhatsApp into a kind of mobile commerce engine to let shoppers sign up for alerts about new products or special offers.

Next up, there were promising innovations like Siri redefining the IT landscape at the intersection of the Internet and mobility. Siri wasn't merely a voice command system; it was a semi-intelligent interactive assistant that allowed mobile users search the web. It could search Google and could even bypass Google's search algorithms for many queries in the wake of specialized search services like Yelp and Wolfram Alpha. Over time, Siri, the revolutionary user interface launched on Apple's iPhone 4S, could become a new face of voice-activated search. Moreover, this futuristic virtual assistant could help mobile users understand social contexts, learn and execute their preferences for specific tasks, and recommend services they liked or appreciated.

Apple's Siri and Google's Voice One were cloud applications, and they provided another powerful venue to reinvent mobile computing via the integration of voice recognition software into the larger mobile Internet whole. Another cloud-centric premise rapidly assimilating the power of the mobile Internet related to location-aware services. Global positioning system (GPS), along with cellular- and Wi-Fi-based triangulation, had been a key building block of the mobile Internet from the start. The GPS sensor was one of the most coveted parts of the smartphone anatomy after the camera; it opened the floodgates of technology innovation through a vast amount of location data and changed the face of mobile Internet. The advent of location- and context-aware technologies had led to services such as Foursquare, Yelp and Nearby that made social discovery an intrinsic part of the mobile Internet phenomenon.

Foursquare was initially a popular social discovery tool for city slickers and allowed mobile users to share their location with friends by checking in via a smartphone app. However, Foursquare subsequently turned itself into a full-fledged location-based social network, a Wikipedia for places. A number of Internet companies—including Instagram, Path and Vine—relied on Foursquare's mapping data to manage the location part of the service. Foursquare's 30 million plus mobile users were able to synch with MasterCard and Visa credit cards and receive targeted offers and discounts from merchants like fast food chain Burger King. The rise of firms like Foursquare was testament of how the location layer of the Internet was transforming the mobile landscape.

Another important case study regarding the connectivity-driven dimensions tied to the mobile Internet phenomenon was Nokia's LiveSight, a bundle of technologies that helped location apps detect buildings that were being viewed through the phone camera. The augmented reality-enabled intuitive mechanism filtered points of interest to only show those in line of sight and froze the camera frame to inspect the location without having to hold the camera pointed at the target. The LiveSight 3D mapping technology was first being employed in Nokia's City Lens augmented reality app which enabled users to point the camera at real-world objects and see data overlaid on top of image on the mobile screen. Pointing the camera at a restaurant, for instance, pulled up online reviews for it. That way the City Lens app put augmented reality to good use and superimposed points of interest on mobile users' surroundings.

Augmented reality was also the primary theme in the next-generation mobile browsers like Layar which allowed users to view their environment and set real-time points-of-interest from the physical world. Layar was a Dutch company founded in 2009 by Raimo van der Klein, Claire Boonstra and Maarten Lens-FitzGerald in Amsterdam. They created a mobile browser that allowed users to find various items by employing augmented reality technology. The augmented reality app placed data from the web on top of the camera view of the physical world and thus provided a mixed version of reality. The Dutch upstart was further expanding the platform by allowing developers to incorporate 3D objects and assign actions such as sounds or clicking on the object for the link.

Fast forward to connected wearables, where augmented reality could be implemented to monitor and give quick feedback on daily life needs tied to, for instance, personal health. The sensor-centric consumer products such as Google Glass were now becoming available, and they were smart enough in sensing and relaying information about the persons who wore them or the physical environments they inhabited. The innovations like City Lens and Google Glass showed how an amalgam of GPS, camera and augmented reality was taking mobile Internet to a whole new plateau. Moreover, the fact that GPS was a core element of the Internet wearables like Google Glass showed how important it was for smart wearable devices to sense objects around them and produce data on user's location, presence, etc.

That also showed how mobile, wearable and embedded computing paradigms were tied together in a larger arrangement called the Internet of Things, and that this vast technology ecosystem was reinforced with powerful technology phenomena like artificial intelligence and cloud-based information storage and sharing. As a result, the Internet of Things was gradually turning into an immersive, invisible, ambient networked computing environment built through the continued proliferation of smart sensors, built-in cameras, software, databases, and massive data centers. The Internet of Things would continue to become pervasive over time and emerging platforms like artificial intelligence and big data would continue to enhance the near ubiquity of this twenty-first century network.

INTERNET OF EVERYTHING

The notion of ubiquitous or pervasive computing was the ultimate dream of techies in the wireless world. One of the most powerful features of mobile Internet was its premise of being always on, and now in this always-on connected ecosystem, almost everything got connected. Moreover, these devices were made smarter with context-based data and software programs, which in turn, would enable them to make human lives easier, more interesting and more efficient. In 2014, the proliferation of mobile technology built nearly into every aspect of computing was taking the industry to the next logical step. While the next phase of mobile Internet evolution was getting used to the intimacy of an embedded user interface a.k.a. the Internet of Things, ironically, the Internet of Things was itself on a fast-track evolution from being connected to just a few things to nearly everything.

That era of pervasive computing was steadily becoming a commercial norm during the mid-2010s. At his 2014 Consumer Electronics Show (CES) keynote, Cisco chief executive John Chambers coined the phrase "the Internet of Everything" for this new era of computing and called it a US$19 trillion business opportunity. Cisco also came out with the number of devices it thought were connected to the Internet by the year 2012: nearly 8.7 billion. However, the top Internet equipment maker acknowledged that the majority of these devices were the desktop computers, laptops, tablets and mobile phones. A smaller number of devices like sensors and actuators represented the emerging Internet of Things camp.

Next, a Cisco spokesperson forecasted that, by the end of the decade, there would be a nearly nine-fold increase in the volume of devices on the Internet of Things. In her blog, Cisco vice president of Corporate Communications Karen Tillman actually put Cisco's estimate at 50 billion devices by the end of the decade. That affirmed the earlier prognosis from Ericsson's chief about the billions of smart sensors and devices to be carried or embedded in networked networks spread across the world—commonly known as the Internet of Things. Morgan Stanley had taken Cisco's data and predicted an even a higher figure: 75 billion connected devices. These numbers had a "network" disruption written all over them.

Two peculiar industry developments hinted toward the coming bandwidth storm. First, in 2012, Google rolled out amazingly fast and relatively affordable fiber optic Internet service in Kansas City. The web services giant also announced to expand this mega-speed network to 34 cities of the United States. Second, in 2013, Verizon Communications paid US$130 billion for Vodafone's 45 percent stake in Verizon Wireless to take complete control of its mobile business and run both wireline and wireless businesses under the same roof. The deal would dissolve the venture the two companies formed in April 2000, which was one of the initial partnerships for the newly launched Verizon, created as a result of the merger of Bell Atlantic and GTE.

Apparently, the future growth of the wireless industry seemed to be associated with the "Internet of Things" hooked up to cars, wearable appliances and household objects. Some of these things would connect to wired home networks, some of them would be linked to mobile networks, and some of them would need to connect to both. The wireless industry had been trying to break bandwidth barriers through ambitious projects that were part of the 4G and 5G enterprises. However, capacity and interference stumbling blocks wouldn't go away anytime soon and wireless wouldn't be able to provide as much bandwidth as wired connections. So, fatter fiber pipes would most likely complement wireless links in this larger network juggernaut for providing near ubiquitous Internet connections for billions of people and devices.

DIGITAL SIXTH SENSE | 171

In 1945, Vannevar Bush presented the concept of memex, a theoretical machine that organized information using multiple screens and cameras. "Consider a future device for individual use, which is a sort of mechanized private file and library. It needs a name, and to coin one at random, ``memex" will do. A memex is a device in which an individual stores all his books, records, and communications, and which is mechanized so that it may be consulted with exceeding speed and flexibility. It is an enlarged intimate supplement to his memory."
Image: *LIFE*

In 1993, Thad Starner, second from the right, built his own wearable computer with a head-mount display. Seventeen years later, he became the technical lead of the epitome of wearable chick: Google Glass. "If you can actually make technology that is not about the interface, but is an extension of your body, that is when mobile computing gets interesting," he told *Atlanta* magazine. Starner also claimed to have coined the term "augmented reality." Image credit: College of Computing at Georgia Tech

7 TWENTY FIRST CENTURY NETWORK

> "We will have many devices that are constantly connected; in that sense, it's connectivity that will be mobile and the devices will merely plug in. This will lead to a world that is not only connected but also live and immediate."
> — Jeff Jarvis, professor at City University of New York

In February 2014, Facebook chief executive Mark Zuckerberg stood in front of the Mobile World Congress crowd in Barcelona to announce the Internet.org initiative, which aimed to cut the cost of delivering Internet services on mobile phones, particularly in the developing countries. The founding members of Internet.org—Facebook, Ericsson, MediaTek, Nokia, Opera, Qualcomm and Samsung—didn't just aim to make access to the mobile Internet cheap in the developing world; they wanted to make baseline connectivity free. The affordable mobile Internet access would in turn help create these companies new business models and scale up information services.

In 2014, about one of every seven people in the world used Facebook, and now the world's most popular social media service wanted to make a play for the rest of people. Zuckerberg called Internet.org a bid to connect to the next five billion people. He had coined the term "on-ramp to the Internet" for a set of core services like messaging, social networking and search. Zuckerberg told the crowd that nearly 80 percent of the world population lived within areas covered by 2G or 3G mobile networks. The Facebook-led coalition Internet.org apparently wanted to bring this mobile phone carrying population into the Internet fold by facilitating the basic Internet services such as search, social media, and e-mail. Here, Facebook had turned to Ericsson, the top wireless network equipment maker, to understand the anatomy of the mobile Internet infrastructure across the world.

"Universal Internet access will be the next great industrial revolution," said Nokia's CEO Stephen Elop at the launch of Internet.org. Nearly a month after the unveiling of Zuckerberg's ambitious Internet.org project to bring mobile Internet to the two-thirds of the world's population, Facebook announced it was buying Ascenta. It was a small British company whose founders had helped to create early versions of an unmanned solar-powered drone, the Zephyr, which flew for two weeks in July 2010 and broke the world record for time aloft. Facebook had brought this small outfit based in southwestern England in less than US$20 million. Ascenta had launched its first drone at the Special Operations Focus Industry Conference in Florida in April 2013. At the time, the company went by the name of "High Altitude Design Ltd" and Ascenta was the name of its aircraft.

The solar-powered high-altitude long-endurance (HALE) drones flew 65,000 feet or 12 miles above the earth's surface, far above jetliners and ever-changing weather. Facebook was going to give wings to the Internet through these drones which were originally designed for border surveillance, anti-poaching, communications intercept or private communications. They served the same function as satellites, but they were cheaper and flew much lower. In remote places with a low population dispersed over wide areas, Facebook planned to beam down Internet connectivity from low-earth orbit satellites, while in denser locations, such as towns, villages and suburbs, it would station high-altitude solar-powered planes circling overhead for months at a time.

The Menlo Park, California–based social Internet pioneer wanted to explore whether access to the Internet could be delivered more cheaply through both new types of satellites and unmanned aircrafts. Facebook announced it was creating a new lab of up to 50 aeronautics experts and space scientists to figure out how to beam Internet access down from solar-powered drones and other "connectivity aircrafts." Apparently, the firm wanted to connect the rest of the world population to the Internet that was otherwise difficult to reach with traditional wired and wireless network infrastructure. Facebook also envisioned that drones could stay aloft for months, and to make the network more efficient, the planes would transmit data to each other using lasers before finally sending it back down to the earth.

INTERNET IN THE SKY

The companies like Facebook and Google were primarily in the business of running web services and delivering ads on those services. If these Internet dynamos were to keep expanding, ultimately, the growth would come from the spread of the Internet itself. So these deep-pocketed firms could not afford to solely rely on the old-school ISPs of the world to carry out the Internet expansion. Facebook was envisioning to give wings to the Internet through satellites and drones, and its much bigger Silicon Valley rival, Google, was trying to bring the Internet to the middle of nowhere through a network of high-flying balloons. While Facebook wanted to place Internet drones over specific population centers, Google aimed to blanket the sky with free-floating Internet balloons.

It would be worthwhile to note here that there was a history of failed attempts to provide the aerial Internet access. During the

1990s, Teledesic, a venture funded by industry luminaries such as Bill Gates, Paul Allen and Saudi prince Al-Waleed bin Talal, had got a lot of people interested but was scrapped even before it launched its first satellite constellation. One problem with these early efforts to bring Internet to the sky was that people on the ground required bulky, custom handsets in order to receive the signal. Things had changed a lot since then.

The rationale behind the airborne Internet projects was now centered on the idea of bringing Internet services to people in undeveloped areas, the so-called "next billion" who would join the Internet mainstream and subsequently rise into the middle class. In a way, these efforts seemed a bit like pie-in-the-sky, but they were coming from two of the most astute technology companies on the planet. These icons of Silicon Valley innovation believed in the high-risk, high-reward approach; Google-backed self-driving cars were a classical example of such technology endeavors.

As briefly mentioned in the preceding chapter, Facebook's chief nemesis Google was obsessed with fixing the world's Internet problem. The digital media stalwart firmly believed that high-speed Internet was the electricity of the twenty first century. The company, while moving forward with its Google Fiber project, also wanted to reach out to the sky and beam Internet signals to the underserved areas across the world through a fleet of balloons. In 2011, Google started testing plastic balloons to bring the Internet to the hinterlands. On June 15, 2013, after two years of development, Google unveiled Project Loon at a press conference in Christchurch, New Zealand. When Prime Minister John Key spoke that day, a few of the thirty antenna-equipped balloons were floating over the Pacific and supplied a tiny and temporary bit of Internet access to some fifty local families.

The goal of this ambitious project was literally to blanket the sky with Internet radios. The solar-powered, high-pressure balloons would circle the globe in rings, connecting wirelessly to the Internet via a handful of ground stations, and pass Internet signals to one another in a daisy chain. Each plastic balloon floating nearly 11 miles above earth would act as a wireless Internet station for an area of about 25 miles in diameter below it, using a variant of Wi-Fi technology to provide broadband Internet to anyone with a Google-compatible antenna. Each balloon had a small Wi-Fi transmitter that could communicate with Internet-enabled devices on the ground. These balloons were being designed to stay aloft for 100 days and they would be able to steer themselves in part by adjusting their altitude, catching atmospheric cross-currents to change direction.

Like Facebook's drones, Google's Project Loon was a radical plan to bring the airborne Internet to parts of the world without infrastructure for web access on the ground. Technologists had been pondering the promise of balloon-based communication for years, but these hot-air balloons were hostages to atmospheric winds. They were much riskier to keep in the air and they moved slowly. Moreover, these balloons were very thin and wore out pretty quickly. On the other hand, however, these balloons were safer and more commonplace. They were also way cheaper than drones. A weather balloon cost a few hundred dollars compared to a drone that cost a few million.

Google engineers were using cutting-edge technologies to cope with the challenge of these balloons getting pushed around by atmospheric winds. For instance, they were using the wind data collected during commercial flights to refine the prediction models, which in turn, allowed them to forecast balloon trajectories in advance. Next, the pump that moved air in or out of the balloon was made nearly three times more efficient, making it possible to change altitudes more rapidly to quickly catch winds going in different directions.

In contrast, Facebook's airborne Internet made a much more rigid network with drones hovering over pre-defined population areas. Drones were more reliable when it came to flying longer distance due to their low consumption of battery power. They were also more likely to be able to maintain a position. Next, drones were much easier to bring in for a landing when there was a need to make repairs or update the software. Google seemed to have acknowledged the merits of the Internet of drones when it acquired the New Mexico-based Titan Aerospace which developed high-flying solar powered drones. Google's Titan buy could complement its balloons-based Project Loon as well as advance its imagery power on Google Maps and Google Earth.

However, as Mark Zuckerberg acknowledged while talking about the Internet.org initiative, the drone-centric high-tech endeavor was in the experimental phase and needed a lot more research. And so did Google's Project Loon which aimed to bring the Internet access to remote locations through balloons mounted high in the sky. Both Facebook and Google had no official launch dates of their respective projects. They were merely shooting for the strategic high ground, and the aim was to bypass mobile carriers and bring large chunks of the population in the Internet fold. These ambitious projects marked the quest for the near ubiquitous global Internet.

Mobile phones—most of them not hooked onto the Internet by the mid-2010s—had already made an amazing impact in bridging the digital divide. Now the race for the airborne Internet was gradually picking up pace and the fact that two of the most successful Internet companies were steering these ambitious technology undertakings made them far more promising. Their deep pockets and innovation track record made the whole idea of Internet-in-the-sky look within reach. These projects were also a classical example of how titans of new tech establishment were setting the trends and coming over the top to win margins and consumer mindshare.

On the other hand, more traditional wireless industry players like AT&T and Verizon were not sitting on their laurels; they were proactively fighting their own battle for a stake in the mobile Internet riches. Mobile phone networks had provided the core platform for the development and operation of mobile Internet which had steadily evolved with these networks over the years. The mobile phone was primary connection tool for most people in the world, and as of the mid-2010s, mobile operators dominated this turf. These companies were also in a better position to create the economies of scale for people at the bottom of the economic system who needed cheaper communication tools. The rest of this chapter will provide a closer look at the evolution of mobile Internet on three generations of mobile networks steered by mobile phone carriers.

MOBILE NET IS THE COMPUTER

There were several keys to the fulfillment of the mobile Internet paragon: the evolution of user terminals and software tools being the prime ones. The next biggest barrier was speed. Sending data to and from mobile devices was initially a tedious process. Speed peaked at around 14.4 Kbit/s, a rate that didn't allow much beyond transmitting e-mails and checking websites for small amounts of information. A lot was at stake for faster wireless data networks. For instance, the future of mobile commerce was intertwined with the realization of fatter network pipes.

Mobile commerce could become a substantial source of revenue for wireless companies once speedier networks were built out. The capabilities of faster mobile networks would also make mobile applications such as messaging and web browsing more appealing to users and bolster new ones such as music downloads, streaming audio and video, and mobile gaming. Although these activities would chiefly be used for non-commerce applications, they would give mobile commerce a boost, nonetheless.

Back in 1999, surfing the Internet on a mobile phone emerged as a very cool idea. However, a couple of years later, waiting several minutes to peer at a few lines of text on a small monochrome screen didn't feel cool at all. Sluggish mobile networks and clunky technology virtually snuffed out consumers' enthusiasm for the wireless Internet. The industry leaders were concerned that users were fed up with overly hyped wireless Internet services that hadn't met expectations. But then 2.5G networks like GSM off-shoot General Packet radio Service (GPRS) came to rescue the wireless industry.

The GPRS spec had tactfully built itself on top of a successful technology, GSM, while it also adapted to the rapidly changing market needs. GPRS—widely quoted as the 2.5G wireless network—represented the first implementation of packet switching within GSM, which was essentially a circuit switching technology. The timing could not have been more perfect for GPRS as it quickly became synonymous with the celebrated premise of mobile Internet. GPRS, with its always-on connectivity to the mobile Internet, eventually became a key landmark in the path toward making broadband mobile Internet widespread.

The work on GPRS started in 1994, and air interface protocol was proposed after an effort that spanned a period of two years. Ericsson signed the world's first GPRS contract with Deutsche Telekom of Germany in 1999. The network, composed of a switching node and a router, directed data packets to the Internet or any other packet-switched data network. GPRS was going to support almost all the major data communications protocols, including IP, which would enable mobile subscribers to connect directly to any data source.

As a GSM extension, GPRS used the same channel, same modulation scheme and the same network backbone to offer data-over-cellular services. However, it utilized packet transfer and routing mechanism and optimized airtime by allocating temporary resources for bursty data applications such as Internet access. The IP interface and support for connectionless services attracted special favor in the wake of the rising tide of the Internet usage. GPRS promised to handle data rates flexibly, according to network availability, from modest 9.6 Kbit/s to as much as 144 Kbit/s, and allowed simultaneous voice calls while sending and receiving data.

The GPRS network offered higher transmission speed and constant connectivity with minimal use of network resources. And with no dial-up, no drop-off, and speeds of 20 Kbit/s to 40 Kbit/s, it provided a testing ground for the mobile Internet services. Inevitably, the GSM industry started to envision an Internet revival through GPRS networks because an always-on connection coupled with data speeds two to three times faster than before could turn wireless Internet into a viable experience. It also allowed wireless operators to demonstrate that they could market mobile Internet services.

"We still don't know how to use these network capabilities to generate new revenues," acknowledged a spokesman of SK Telecom, the first company in the world to launch commercial 2.5G service in October 2000. SK Telecom, which commanded half of South Korea's mobile subscribers, made the mistake of jumping onto the new-age wireless bandwagon without proper groundwork and was now scrambling to recover. For a country in the vanguard of hot wireless technology, nobody was able to figure out what's going to be on offer in the 2.5G platform. A year since the commercial launch of its 2.5G service, to the frustration of SK Telecom managers, the subscribers were simply talking up against their ears in the traditional fashion.

SK Telecom's dilemma marked the crucial importance of another component that complemented the network: content. Eventually, it was the world's largest content database—the Internet—which came to rescue the promise of fatter wireless pipes like 2.5G networks. The Internet's first killer app, e-mail, became a must-have item for business executives and mobile workers across the United States. BlackBerry, the awkward cousin of the PDA lineup, allowed people to wirelessly collect e-mails and reply on tiny but workable built-in keyboard. So the knee-jerk phase faced by the likes of SK Telecom passed rather quickly.

BlackBerry—the first truly successful smartphone—played a significant part in stimulating the early demand for GPRS networks. The Waterloo, Canada–based firm BlackBerry was a pioneer in the paging and wireless messaging business. The early BlackBerry device introduced in 1999 communicated using the low-speed Mobitex network, the one that also provided refuge to the trivial Palm VII launch. It was the launch of the 2.5G mobile data technology that delivered an always-on connection and allowed data-centric gadgets like BlackBerry the ability to jack into the Internet with a faster e-mail connection. BlackBerry reached a wider market in the early 2000s with the launch of GPRS-supported phones such as BlackBerry 5810 and BlackBerry 6210.

The Internet largely took care of the content conundrum on mobile platforms. However, the wireless industry faced other crucial stumbling blocks. For instance, the voice-centric wireless communication world was now operating in a new territory—mobile Internet—and it wasn't an easy turf. Wireless operators had built a great intelligence on voice usage over the decades, but for data, the infrastructure and efforts were generally not on par. Initially, there was little understanding of what consumers were doing, and which applications and services they were tuned to at any given instant. The cost of supporting data services exceeded the cost of managing voice services and the revenues from data services only became prominent after the arrival of the iPhone.

With the iPhone and credible rivals like Android phones came a growing appetite for data—first in the browser space and later with mobile apps. Case in point: the iPhone was the primary driver behind the unprecedented 5,000 percent increase in data usage on AT&T's network from 2007 to 2010. The complete Internet experience that people have had on PC was now available in the mobile space, and that was nothing short of a revolution. The union of ubiquitous high-speed wireless networks with mobile browsers meant people had the power of the web in the palm of their hand.

Here, at this technology crossroads, the advent of the iPhone was also a stark reminder of the critical need for greater network speed and capacity for a robust mobile Internet operation. The so-called 2.5G networks provided decent-enough data rates for WAP-based services, but as cell phone browsers matured beyond simple, text-heavy displays, the need arose for a bona fide mobile network capable of handling high-speed data rates. The advent of smartphones and mobile Internet had underscored the need for a specialized network, and there was no looking back. Smartphones, more than any other device, were able to take advantage of the faster networks to empower users to do or access nearly anything.

The correlation between the mobile Internet and the third-generation (3G) wireless networks was evident in the post-iPhone wireless world. By 2009, AT&T had been turning heads with the sheer speed of its 3G network. The mobile phone industry now had a challenge of wirelessly connecting millions of users to the Internet, and for that, they needed far more capacity than required in providing plain voice services. Smartphones were now into the mainstream, and for this new value proposition, the wireless establishment needed to have faster networks so that they could manage the onslaught of this new gadgetry. The next generation of wireless networks was on its way.

A SPECIALIZED DATA NETWORK

The early promises of 3G had created a marketing field of dreams in which video would vault mobile phones into an entirely new plane just like the addition of pictures to sound in the new medium of television had transformed broadcasting in the 1950s. The most compelling dimension of 3G, however, was the one that related to the build-out of the next-generation wireless infrastructure for the mobile Internet.

When European planners in Brussels were mapping out the timetable for 3G launch, across the Atlantic, the Internet was moving into the consumer realism and web startups like Netscape had become stock market darlings. Understandably, while piecing together the 3G prototype, Europeans became convinced that the rapid coalescence of mobility and the Internet would crystallize a new market with a powerful and timely convergence. A decade ago, similar central planning for GSM development had helped transform Europe into the world's richest wireless market, turning the regional wireless players, from Nokia to Vodafone, into global powerhouses. So while creating a new continental network—3G—high-speed Internet became Europe's audacious bid to lead the world in a crucial twenty-first-century industry.

For Europeans, it was now a megaproject, the equivalent in size, vision, and expense of America's Apollo space program in the 1960s. This was a unique blend of technological conquest and pervasive market drive borne out of Europe's GSM triumph. The pundits called 3G wireless the vision of the new century, a wireless nirvana where all dreams would come true. The so-called phone of the future riding on 3G platform would be powerful enough to provide high-speed Internet access, video-on-demand and countless whiz-bang features. According to early projections, 3G was to provide users with a whopping 2 Mbit/s of data, more than what most wired offices used to connect their users to the web at that time.

However, 3G wireless was likely to cost an operator several billion dollars and many years to roll out a complete network. Plus, 3G coverage could remain spotty and sporadic for years. But despite setbacks, on both technological and marketing fronts, wireless industry's commitment to deploy 3G remained unshaken. The wireless companies picked up the pieces and began to work out the kinks for providing quality phones that would subsequently accompany this new technology. Once 3G began to take hold, cell phone companies worked tirelessly to find ways to utilize this new technology. After initial alarm bells, by early 2000s, network construction was well under way, albeit at a slower pace than predicted earlier.

Buying into 3G would mean buying a promise: wait a few more years, and buying the real thing would become more likely. That real thing came by in the late 2000s when the iPhone and Android pushed the mobile Internet well into the mainstream with hundreds of millions of subscribers. Although mobile phones long had the ability to access data networks such as the Internet, it was not until the widespread availability of good-quality 3G coverage in the late 2000s that specialized devices like the iPhone emerged to access the mobile Internet. As coffee-shop goers and technology experts alike attested the speed was completely real.

The wireless industry—then focused on voice and messaging services—was caught unguarded by the explosive growth in mobile Internet traffic. For them, data was only useful to the mobile enterprise workers and early adopters who used smartphones with clunky browsers that made mobile web surfing far less appealing. But Apple's seductive phone with its powerful software and intuitive interface encouraged users to go online and stay online. The iPhone users consumed an average of up to ten times the bandwidth of mobile subscribers. They were playing games on their phones, sending video messages, and downloading music. A new network regime was taking shape in a wireless order created by smartphones like the iPhone.

Still, the 3G-based mobile Internet had turned into one big traffic jam, and unlike wired Internet, where more bandwidth meant laying more fiber cables and adding more servers, there was only so much bandwidth that mobile carriers could squeeze form the available radio spectrum. Not surprisingly, therefore, AT&T discontinued unlimited data provision and migrated to the capped data plans only two years after launching the iPhone in 2007. Verizon's unlimited data plans only lasted for a few months following its iPhone launch in 2011.

The iPhone had created the consumerization of mobile data services, but on platforms like the mobile Internet, consumerization was generally accompanied by a catastrophic decline in per bit revenue for the mobile service provider. Furthermore, the more iPhones were sold, the more unhappy users could be because of the network congestion. The world had witnessed how Apple's iPhone crippled swaths of AT&T's wireless network as users complained about dropped calls and network delays. The network that pioneered the smartphone in collaboration with the iPhone also topped in consumer dissatisfaction.

Around the world, mobile operators such as O2 in Britain and SingTel in Singapore were facing similar struggles. Moreover, by 2010, with the availability of media savvy devices such as iPad hooked onto mobile networks, it had become evident that, at some point, 3G networks would be overwhelmed by the growth of bandwidth-intensive applications like video streaming. So the wireless industry began looking for data-optimized technologies with the promise of speed improvements up to tenfold over the existing 3G technologies.

THE REAL MOBILE INTERNET

Internet downloads and GPS position finding could be accomplished with 3G speeds but *Dick Tracy*-style video calling could not, at least not without significant compression techniques. While video chat and media consumption were seen to be two key use cases on the upcoming smart devices, streaming a single video from the web to a mobile device took as much bandwidth as ten phone calls. Then there were applications like music and games that continued to grow at a rapid pace, making it imperative for the wireless establishment to ensure that bandwidth stays ahead of customer demand, now and in the future. Cisco Systems had predicted that worldwide mobile data traffic would increase thirteen-fold between 2012 and 2017.

The average data usage per device was on the rise, and so was the total number of connected devices each person owned. Moreover, not only were the number of smartphones and tablets increasing, but more devices from cameras to cars were getting connected to the mobile juggernaut. As the saying goes "out with the old, in with the new," such was the case with 3G networks gradually making way for mobile Internet-centric fourth-generation wireless systems commonly known as 4G, the network that promised to be faster, stronger, and overall better than its predecessor. Wireless companies hoped that 4G would revolutionize the way people connect to the Internet. They even anticipated that 4G would partly eliminate the need for smartphones to search for Wi-Fi hotspots in order to connect to the high-speed Internet.

Long Term Evolution (LTE)-based 4G technology promised speeds that were about ten times faster than they were on 3G networks. And like its 3G predecessor, the industry gurus started calling it a game-changer the sooner 4G made its name on the wireless paraphernalia. But a closer look at this generation game revealed that the labels like 2G, 3G, and 4G didn't really matter much. Advances were being made and would continue to be made, providing consumers with better, faster mobile broadband, whatever generation it was, and that's what mattered.

There were distinctions, however. One of the key ways in which 4G differed technologically from 3G was in its elimination of circuit switching technology which was common in traditional telecom settings—instead employing an all-IP network. The crucial significance of 4G was lying in the fact that it aimed to be a full-fledged wireless IP network and that it would potentially close the capability gap between wired and wireless Internet worlds. Earlier, the ITU had standardized 3G networks for use on both circuit- and packet-based networks, meaning that voice would travel across the telecom-based network and data would be transferred through the Internet-based network. Conversely, 4G—being the first generation of wireless network technology which was completely IP-based—was set up to use the packet technology only.

The general idea behind 4G was to provide a comprehensive and secure all-IP based solution, which in turn, would facilitate IP telephony, ultra-broadband Internet access, gaming services and multimedia streaming to mobile users. With an all-IP makeup, 4G would allow a treatment of voice calls just like any other type of streaming audio media, thus providing a key turning point for smartphones, which were now increasingly comparable to personal computers.

Long Term Evolution or LTE, as the name suggested, was actually evolution of the existing 3G standard, not a new standard in its own right. It was a prodigy of the classic GSM technology evolution path. In fact, LTE was long-term evolution of the UMTS technology, and it had built itself on 3GPP technologies like GSM, GPRS, EDGE, W-CDMA, and HSPA. In other words, LTE was simply an advanced form of 3G, also called 3.9G by the ITU. However, LTE represented a paradigm shift from hybrid voice and data networks to data-only networks. The network architecture for LTE was greatly simplified from its predecessors because it was a packet-switched network; it didn't have the capability to handle voice calls and text messages natively.

The advanced version of LTE would offer a theoretical capacity of up to 100 Mbit/s in the downlink and 50 Mbit/s in the uplink, and more if MIMO technique for antenna arrays was used. The relentless march of bandwidth had started with the ascent of smartphones, and it would only go farther. People had started streaming movies from the web to their smartphones and iPads, and as they did so, the demand for mobile bandwidth grew faster than anyone ever imagined. So it became imperative that mobile phone operators figured out how to slice and dice data into an appealing tiered plan because once 4G network deployments passed halfway mark, there would be no looking back. Wireless operators were going to become mobile Internet service providers (ISPs) with voice business on the side.

Here, wireless operators didn't want to fight the mobile Internet battle just to become dumb pipes. The flip side of the mobile Internet juggernaut for the network carriers like AT&T was that they had a lot more at stake than did the newcomers like Apple, Facebook and Google. AT&T's former chief Ed Whitacre had raised the alarm bells back in mid-2000s by saying that the Internet companies like Google were getting rich off his network pipes. The Internet was changing, and the mobile business was changing. At this gigantic crossroads of the Internet and mobility, the mobile phone operators were adjusting to a new power dynamic in which the "big four" of the new tech establishment—Amazon, Apple, Facebook and Google—were setting the trends and coming over the top to eat their margins and consumer mindshare.

Amazon with its US$34 billion online retail operation and Facebook with nearly a billion-strong user base could make a play for the mobile Internet pie. Apple and Google were already leading the charge in this realm with an active culture of innovation. In this particular backdrop, the next chapter picks a specific industry in the wireless domain—mobile commerce—and takes a good look at it as a case study that is intertwined with the overall development of mobile Internet technology and business. Mobile commerce was also a crucial market as being the common thread in the future roadmaps of the four horsemen of the mobile Internet: Amazon, Apple, Facebook and Google.

196 | THE NEXT WEB OF 50 BILLION DEVICES

Norwegian mobile operator NetCom's staff members accompanied by Huawei engineers set up the world's first live mobile broadband Internet link based on a 4G connection on June 5, 2019. Later, on December 14, 2009, TeliaSonera and its Norway-based subsidiary NetCom launched the first 4G service over an LTE network in Oslo. Huawei Technologies supplied the network equipment for this launch.
Photo credit: 3g.co.uk

TWENTY FIRST CENTURY NETWORK | 197

Facebook envisioned to set up an Internet of drones and to bring web services to parts of the world that were still offline.
Image courtesy of Facebook

8 MOBILE COMMERCE: A CASE STUDY

"In my opinion, the 'future of mobile' is the 'future of everything.'"
— Matt Galligan, co-founder of SimpleGeo, a location services company

In 2007, soon after Olli-Pekka Kallasvuo took charge of Nokia as the new CEO, he embarked on an ambitious technology acquisition. Nokia bought the Chicago-based digital map company Navteq Corp. for hefty US$8.1 billion to bring navigation out of the car and deliver it to pedestrians. Nokia coined the buzzword "context-aware Internet" while asserting that it would reshape the web. To accomplish that Internet panacea, the Finnish mobile giant was pinning its hopes on operator-independent, cross-platform phones conceived through the development of new software and services. The company claimed that the Map

2.0 would enable context-aware Internet by combining multimedia features with the freewheeling Internet and assisted-GPS technology.

Nokia engineers asserted that by adding context—such as time, place, and people—to the Internet, the mobile web experience would become something entirely different. Once the context was added to the network, they contended, the Internet experience would become more mobile, contextual, and personal than on the desktop. The building blocks necessary to make this happen included GPS, broadband wireless access, a back-end service, and enough processing power and memory residing on mobile phones. Here is one scenario depicting how it would actually work: a user takes pictures with a camera phone, and the GPS coordinates are simultaneously stored in a metadata file; Bluetooth could sniff around and discover who is around him or her. Location, therefore, would no longer be an application; it would become core fabric of the mobile Internet.

Nokia managers loved to play up fascinating new scenarios at technology press events. Their hyperbole was reminiscent to the early days of mobile commerce talk which was stimulated by the arrival of WAP-based mobile phones back in the early 2000s. At that time, marketing dream weavers conjured up whiz-bang scenarios in which mobile-phone users would resort to all kinds of amazing adventures. One might have heard this: walking down the street, a user approaches a Starbucks coffee house and his or her mobile phone starts ringing; on the handset screen pops up a coupon for a $1 latte. Or this: A user strides into a department store and slips into that perfect pair of jeans. A bit pricey! No sweat for his or her mobile phone. The user punches the barcode of the jeans into his or her handset and receives 20 percent discount from an online retailer.

However, these marketing potions carried a fundamental flaw: the key building blocks to make mobile commerce a commercial reality were not ready yet. What Nokia did in 2007 was recycle this notion by combining two chic technologies of the time—the mobile web and GPS-based location—and started spreading the context-aware Internet gospel. But after three years and some failed projects, there was little evidence of any tangible payback to Nokia's foray into location-centric premium phones. When the dust settled, Nokia seemed to have fallen to the classical marketing paradox that was all too familiar in the twilight world of the Internet and mobile phone.

Fast forward to 2010, and it was a very different story in the post-iPhone arena. Hundreds of millions of smartphone owners were using mapping apps like free turn-by-turn navigation; location was also bubbling away as a critical feature in social networks like Facebook, Foursquare, and Twitter. Maps and location became so critical on mobile devices that it would be considered crazy to have a smartphone that didn't include GPS and software providing location-based services. The iPhone users were also able to find a restaurant of their choice and tweet from sports events.

Apple, in a stark contrast to Nokia, carefully rationalized the supporting technology components, worked out a robust product roadmap, and then mobilized its legendary marketing machine. The iPhone was initially conceived as a phone that would play music and video from iTunes, but its primary appeal quickly became the millions of head-slapping, useful software applications that ran on it. The iPhone certainly played music, but owners were just as likely to use it to check the weather, book dinner reservations, read a newspaper, get directions, or play a quick game of Taxiball on the subway. The creation of the App Store for the iPhone was a revolution that literally changed the smartphone world overnight and helped the iPhone reach dizzying heights.

The Cupertino, California–based company sent shockwaves across the wireless world by readily shifting the consumer expectations for smartphones and mobile Internet. The App Store created billions of dollars in new revenue for software makers and catapulted the iPhone into being the largest must-have device on the planet. Suddenly the competition found itself scrambling to make its own version of the App Store, create software tools, and get programmers on-board to begin making applications for their devices. Companies as varied as Amazon, Microsoft, Nokia, Motorola, and BlackBerry rushed to adapt, also adding touchscreen-based devices to their product lineups.

The iPhone was also the first mass market mobile device that made the Internet fun and easy to use, especially on websites optimized for mobile phones. Once phones and the mobile Internet converged on powerful new platforms, like the iPhone, more consumers started using the mobile web as a viable communication tool. Devices like the iPhone boasted significant computing power and made accessing the Internet from a handset far easier than with mainstream feature phones.

The combination of smart handheld devices, mobility-enhanced applications, GPS software, and robust wireless Internet experience added up to the promise of mobility. Apple had successfully demonstrated that those companies that effectively execute mobile strategies for commerce, entertainment, and finance would be the winners in the mobile Internet arena. When the movers and shakers behind m-commerce's false start of early 2000s saw the changing tech headwinds, they rejuvenated themselves to bring users banks, shopping malls, entertainment centers, and all of their friends to their palm, pocket, or purse.

In many ways, the mobile commerce market was a mirror image of the development of mobile Internet; the evolution of these two industries had a lot in common. The mobile commerce marketplace also provided an important case study of the "apps versus web" turf war that the book has broadly chronicled earlier in chapter 4. Apps clearly dominated the mobile commerce scene in the post-iPhone arena, but as the next section will show, the initial hyperbole unmistakably related to the mobile web.

A BRIEF HISTORY OF MOBILE COMMERCE

Mobile e-commerce, or simply m-commerce, harnessed the ability to make commercial transactions from a mobile phone. Back in 1999, in the heady days of dotcom euphoria, it all looked so easy: to make money from the mobile Internet, simply create a mobile version of what worked so well on the fixed-line Internet. The notion of m-commerce quickly became the face of the mobile Internet in a way that was reminiscent to the rise of e-commerce soon after the wired Internet took shape in the commercial arena during the mid-1990s. Then, people would call e-commerce the second chapter of the Internet. Mobile commerce went one step further by promising the ability to purchase goods anywhere through a wireless Internet-enabled device.

Mobile commerce drew a lot of strength from the thriving wireless web premise because of the ubiquity of mobile phones. Surely, anything that could be sold over the conventional Internet to PC users could be sold over cellular networks to mobile subscribers. And because mobile users had their phone with them at all times, they might be expected to do more shopping than stationary customers. With the rapidly maturing concept of the mobile Internet, the early backers reasoned, m-commerce could serve as a powerful platform to host new services in collaboration with Amazons and Yahoo!s of the Internet world.

Here, the mobile commerce story took an interesting twist. Although the Internet euphoria spread across the whole globe, when it came to business realities, it was pretty much an American phenomenon. From the super-ISP America Online to the Internet hardware king Cisco to the UNIX server powerhouse Sun Microsystems, almost every predominant company on the Internet scene was of American origin. Hence, when commercial Internet extended its reach to the big-time e-commerce bonanza, it seemed like just another American show. In the ongoing wireless revolution, however, Europe and, to some extent, Asia called the shots. In countries like Japan, where the personal computer penetration was low due to cultural reasons, mobile phones could well become a potent tool to access the plethora of nifty consumer services.

Inevitably, the companies in these parts of the world thought up of m-commerce as an answer to America's e-commerce hegemony. But although European and Japanese footprints were initially predominant in the m-commerce roadmap through projects like WAP and i-mode, when it came to actually doing the Internet, America seemed to know the way forward. It became evident that the Internet-savvy America had an important role to play in this nascent marketplace when Amazon.com Inc. launched its m-commerce efforts in late 1999. The Seattle–based e-commerce upstart had invented one-click ordering on the Internet that let the buyers store credit card number and address after the purchase. Next up, the online retail pioneer, through its "Amazon Anywhere" initiative, started assembling partnerships with a number of cell phone operators in the United States.

The arrangements typically called for Amazon's website to be given a prominent placement on the screens of mobile phones, critical in those early days because of difficulty users had in navigating the web by punching on a phone keypad. In March 2000, just before the dotcom bubble burst, Amazon chief Jeff Bezos predicted that by 2010, all of his firm's customers would use wireless devices to make purchases. Describing m-commerce as "the most fantastic thing that a time-starved world has ever seen," he predicted that it would change the way people shop since they would be able to make impulse purchases anywhere at any time. Within five to ten years, he claimed, "almost all of e-commerce will be on wireless gadgets." In an industry not yet humbled by the dotcom collapse, Bezos wasn't alone in seeing the wireless web-based mobile commerce as the next big sensation.

Market analysts queued up to make rosy forecasts of m-commerce revenues. With such a bonanza conceivably around the corner, it's no wonder that wireless operators paid so much for the 3G licenses. However, when the Nasdaq crashed and the dotcoms started going under, the wireless world came to a rude awakening that making money was hard enough even on the conventional Internet, where technology was rather mature. Once-fashionable dotcoms had failed to bring a workable business model and were now burning in style. The dotcom fiasco had shown to the world the ugly side of the Internet. Now the prospect of buying things on cell phones, with their tiny screens and keypads, suddenly looked far-fetched. The surveys conducted after the dotcom crash showed that consumers found the reality of m-commerce hugely disappointing.

In the hindsight, the way America Online, Yahoo! and Amazon approached the wireless Internet was to take what they had on the wired web and simply put that on the handset. Another problem with the mobile Internet was that there were too many clicks. An early trial found that it took over forty minutes to order a book by cell phone. Likewise, booking a hotel room on a mobile hotel-reservation system required thirty-seven clicks. Then there were those dreaded lessons from the e-commerce chapter. For instance, its ad-based revenue model had initially proved untenable, leaving many in the wireless industry wondering whether consumers would be the ones to pay for mobile services. Banner ads, pop-ups and the like were deemed failures in the wireless domain. On the heels of this confusing picture, online merchants started placing a low priority on developing and marketing mobile commerce services. The user base also remained small because only a few mobile websites were equipped to accept m-commerce transactions.

MOBILE WEB'S AWKWARD ADOLESCENCE

In a way, mobile commerce was just any wireless data activity that made money for a company along the value chain. So when a mobile-phone user made a restaurant reservation, the amount that the user later paid using cash or credit card could be counted as m-commerce revenue. Under this view, mobile commerce, in collaboration with the mobile Internet drive, could potentially bring a myriad of new activities for the wireless industry. For a common user, it was all matter of value. Once the wireless industry worked out this piece of puzzle, m-commerce and other wireless data services could come of age much faster than anticipated.

How the promise of m-commerce would descend to the cyberspace was still a question mark in the early 2000s. However, as wireless devices got faster, smarter and cheaper, the hope floated that more effective software platform to power the m-commerce products would subsequently emerge. The broad consensus was that it wouldn't be before 2005 that the world sees the three A's of m-commerce—anything, anytime, anywhere. In the meantime, m-commerce architects had to find ways to complement the desktop web but avoid replicating it. There was little point in making an impulse purchase of a book or a CD if it would then have to be delivered by post. Rather than spend ages pecking at a phone keypad, why not wait until one get home and order in comfort from a PC.

The mobile Internet was different from its fixed-line counterpart in three important aspects. First, a mobile phone was a far more personal device than was a PC. It was likely to be used by only one person, who would probably have the phone with him or her for most of his or her waking hours. Second, in the pre-iPhone era, mobile phone operators could broadly determine what menus and services appeared on their handset screens. The ability to set the default portal was a big advantage for operators because it allowed them to act as gatekeepers. Last, and most important, people knew that using mobile phones cost money, and there was a mechanism for the network operator to charge them for that use. Sending an e-mail over the Internet from a PC was essentially free; sending a text message from a cell phone cost about 10 cents. So users were more likely to pay a mobile premium to do things while on the move.

Beyond the stark contrasts of wired and wireless Internets, another significant challenge for the wireless industry was to make mobile commerce a seamless social experience. The traditional Internet had existed for many, many years and was only used by a handful of scientists because it was too complicated. Only after people got reasonably easy way to browse the web did the Internet explode and became a driving force in the economy. The same could happen to the mobile Internet and eventually to mobile commerce. The logical conclusion: make it easy.

Then there was this risk that wireless operators might be tempted to set up "walled gardens" of services and contents, hence restricting users to a handful of approved services that would enable operators to capture a much larger chunk of the expected bonanza in data revenues. In the 1990s, online services such as America Online, CompuServe, and Prodigy operated on the walled-garden principle; however, as soon as one of them offered unfettered Internet access, the others had no choice but to follow suit. The walled-garden model could turn out to be just as unsustainable on the mobile Internet because it annoyed users.

In fact, in the early days of the WAP launch, mobile carriers generally limited number of sites offered to customers by signing exclusive agreements with content providers, or selecting a package of sites from a content aggregator—a third-party portal or a WAP provider. But subscribers didn't want restrictions on whom they could do business with or which sites they could visit. This strategy backfired as the majority of users found the mobile commerce experience unsatisfying. Mobile phone companies subsequently removed these restrictions. Mobile commerce was initially written off also because the early versions of Internet-enabled phones, the so-called WAP phones, didn't deliver what they promised. When WAP failed to live up to expectations, there was a backlash for mobile commerce. In many ways, WAP became the acid test for mobile commerce viability.

Mobile Internet access providers, in this case, the wireless operators, generally charged by usage, either for every minute spent online or for every byte downloaded. This meant that they made money on transporting data, so it made sense to offer users the widest choice of content possible to encourage them to run up transport charges. That's how i-mode worked in Japan; the vast majority of NTT DoCoMo's data revenues came from transport, not the sale of content. The Japanese mobile phone operator generally offered a selection of approved services through its own chosen portal, but also gave subscribers the option of going elsewhere. That's what America Online did with its dial-up Internet service; it offered services such as instant messaging, chat rooms, and e-mail, as well as access to the web.

MOBILE COMMERCE: A CASE STUDY

Wireless industry, primarily made up of traditional telecom service providers, faced an important crossroads. The telecommunications model was predominantly based on a closed architecture. A telecom operating company typically owned the network; provided the transport; supplied the services, such as caller ID and calls waiting; and billed the subscriber directly. But was this model applicable to the convergence of mobility and the Internet? Probably not! The ability to control the entire value chain didn't exist in many industries. Now the convergence of voice and data was testing this model because original expectations of the mobile Internet were largely based on the experience mirroring the wired Internet.

In the open architecture of the Internet, service providers bought the raw bandwidth from a backbone carrier such as UUNet while end users paid for their connection media: dial-up telephone line, DSL, or cable. Once connected, they could subscribe to and pay for any service and complete transaction at their discretion. The consumer accepted this model in which the transmission of data was separate from the provision of service. Overcoming these expectations was proving a challenge for wireless operators.

The industry players were making all the right noises, but because of the complexities of initiating the technology, mobile commerce marketplace simply didn't seem ready in the early 2000s. The key m-commerce players had more than a few months of experience to draw on it. Then there were all these whacky ideas like web-enabled ice cream delivery. While it sounded fascinating to have a refrigerator send users an e-mail to let them know they were out of milk and then send a message to Webvan to bring more, in reality, this only came out as one of the "oh-golly" scenarios.

The Internet strategy had indeed become indispensable for every wireless company, but the problem was that the public seemed unenthused. Mobile phone users at large were not sure what the wireless Internet was because the medium itself was still in its embryonic stage. The slew of companies that once emerged hoping to make it easier to translate the riches of the web into the wireless world also missed one important step. No one bothered to ask consumers what they wanted. The wireline Internet had taken off because it offered users instant gratification to get what they wanted, when they wanted it. Next up, mobility started giving users the additional space to get answers when and where they wanted them.

Back in early 2000s, for the wireless Internet, while much had converged, much also remained to be converged. Although consultancy firms had predicted massive m-commerce revenue streams, there were big hurdles—standards, security and presentation—before these vast revenues could be tapped. Many of the mobile consumers' concerns, such as security and privacy, and difficulty with navigation, were reminiscent of worries of the early days of e-commerce and could eventually be overcome. Still, there were broader problems with using handheld devices for shopping. Compared with PCs, which had large color screens, handheld devices of early 2000s were hopeless for browsing. Scrolling through lists was cumbersome, and features and prices were hard to compare.

But the difference between these two worlds also presented a window of opportunity. People carried mobile phones with them everywhere, so it was only a matter of time before all necessary components converged onto the mobile phone, making it even more invaluable.

THEN CAME THE iPHONE

The mobile Internet's troubled youth was a form of natural progression in the evolution of mobile commerce. Mobile commerce generated a huge buzz when introduced during the late 1990s, but with the exception of NTT DoCoMo's i-mode offering, it had been a major disappointment. Due to network and handset shortcomings and limited content available on WAP and other data services, potential customers were not enthused and failed to sign on in large numbers. Mobile commerce went through a lot in its brief existence and the insight this learning curve garnered from failures of wireless operators, content providers, and WAP service companies proved to be immensely valuable. For instance, the wireless industry began to ponder on why had m-commerce been a hit in Japan only?

One of the primary reasons of WAP failure was that services were launched before they were ready. Moreover, mobile phone networks were inadequate for handling the commerce-like shopping experience that was promised to subscribers. Mobile commerce was a bright idea whose time could come once the wireless industry came over its apathy for embracing data services. However, the basic problem with mobile Internet in general, and mobile commerce in particular, was that there was no clear business model. The remedy would require the transformation of the existing telecom practices into entirely new business models, and it was an uphill task. However, if done well, mobile phones, for instance, could emerge as a popular vehicle of payment, at least for small items. Since wireless operators were used to handling large numbers of small transactions, their billing systems could manage such transactions at around a tenth of the cost of a bank or a credit card company.

Unlike on the fixed-line Internet, people were prepared to pay for content and services they really wanted. But they preferred to pay lumpy subscription fees rather than a small charge for every morsel of information they accessed. Once past the initial stumbling blocks, mobile commerce, being rife with opportunities, could kick-start in several fast-growing niches: for example, provision of services to stock traders and others who need instant information. Mobile commerce could also transform stores into virtual showrooms. Another use case: a person looking for an apartment, while driving in a particular neighborhood, saw a "for rent" sign with a code listed on the signboard. He or she could key in the code into cell phone to receive short video clips of the interior and decide about arranging a visit.

MOBILE COMMERCE: A CASE STUDY | 215

In many ways, the mobile Internet in 2005 was at the same stage of development as its wired counterpart was in 1995 when the web was in its infancy. Then, retailers asked if the e-commerce channel was for real. Could it be used as a competitive channel? There were hundreds of startups, and nobody really knew which technologies and business models would win, or what consumers or corporate users wanted. But this cycle of boom and gloom on the e-commerce trail also meant that there were plenty of lessons to be learned from the mistakes made on the fixed-line Internet. While which model would prove most successful remained to be seen, there was certainly money sloshing around on the mobile Internet.

By the late 2000s, most of the pieces of the puzzle that had to be in place to make m-commerce successful were already in place. Text messaging had provided m-commerce with initial traction and necessary learning curve, but SMS had mostly been successful as an appetizer. That's probably because text messaging was primarily a marketing vehicle; when it came to making purchases, the action was on mobile websites. Only fully transactional m-commerce sites would enable consumers to shop more or less as they would on an e-commerce site, except on a pared-down version. It was also imperative that m-commerce sites provide shoppers access to the same number of products as e-commerce sites.

The search for the "killer application" went on until the iconic iPhone came along, bringing a revolution in the way people look and use their mobile phones. The iPhone launch was a defining moment for the nascent m-commerce industry. After the iPhone-driven commerce gathered initial momentum, many in the industry started taking a more rounded, holistic approach that embraced not only payment apps, but also mobile-optimized sites, mobile advertising, mobile coupons and tickets, and location-based services.

Retailers had heard the m-commerce mantra before. A lot of retailers that had earlier invested resources and effort in the mobile market were disappointed in results, which generally stemmed from consumers' poor experiences on conventional mobile phones. Also, in the early stages, mobile-optimized sites were quite dumbed-down; they were basic versions of websites that reminded users of clunky web pages of the early 1990s. And the fully capable mobile Internet devices were not there yet. The iPhone changed all that.

The slick new phones with great visual experiences and added functionality like GPS location awareness changed everything. Just as broadband Internet made shopping online far more attractive, so too, did these powerful smartphones by changing the m-commerce game. Now retailers were lining up to learn the rules of the m-game. The post-iPhone era saw a flurry of retailers creating text messaging-based marketing programs, m-commerce sites and downloadable mobile applications to reach out to consumers who wanted to do more with their phones. Take eBay, for example, which reported a whopping US$380 million in sales through its iPhone app and m-commerce site for the first nine months of 2009. Likewise, a Nordstrom app replaced a salesperson with a whole new in-store experience.

Wireless companies initially touted m-commerce as a mobile extension of e-commerce instead of portraying it as a unique, value-added mobile service. So they found users flocking to the wired web and ignoring the mobile alternative. Customizing m-commerce to multiple consumer tastes, as shown by DoCoMo's i-mode service, launched nearly eight years before the arrival of iPhone, offered better chance of success. The i-mode service allowed Japanese mobile users to utilize their phones for most of their everyday transactions. They could purchase a plane ticket and use the e-statement stored on the phone to check into the flight. They could locate nearby convenience stores or vending machines using mobile GPS and then make the purchase from the device itself.

Like its remarkable mobile Internet execution, Apple seemed to have learned all the right lessons from the success of i-mode in the m-commerce realm. Like DoCoMo, Apple carefully cultivated an entire ecosystem made up of thousands of apps, which in turn, made it possible for people to go beyond the mobile web to get things done. The company that had put MP3 players into the hands of the masses and had changed the way consumers interacted with music was now courting software developers to build a brand new industry around its iPhone platform. By 2010, the iPhone was at the center of a huge m-commerce maelstrom with thousands of alluring apps. An increasing number of consumers armed with smartphones like the iPhone were relying on online product reviews and recommendations while in the aisles of brick-and-mortar retail outlets.

Before the iPhone came along, businesses had one simple question: "What should we be doing on mobile?" In the post-iPhone era, they were unanimous in their agenda: "We should be doing an iPhone app." With apps, the smartphone suddenly became the epicenter of the software world. With an app phone in a user's hand, which was like a mini computer, a revolution was knocking at the door, and it was like the PC fable all over again. Take the case of gaming, which exploded on the iPhone platform in 2009, especially with fantastic shooters such as *Alive-4ever*; with quality game-play and graphics it looked like something seen on an Xbox 360. Then there were apps like UberCab, which allowed consumers to call a taxi. These early apps either served as a utility tool or provided some kind of entertainment. However, as it turned out, there were bigger opportunities hidden in the apps domain.

The next year, in 2010, smartphones started to change the e-commerce equation by enabling shoppers to bring the means of buying online straight from a traditional store. Amazon's Price Check—a free iPhone app—let users check prices of CDs, DVDs, books, and video games on the fly by scanning products, snapping a photo, saying the product name, or typing in the name, brand, or model numbers. Price Check burned away the awkwardness of typing the names of individual items into tiny search boxes by building a barcode scanner into the app. The days when consumers merely crawled through newspaper ads and trekked out to brick-and-mortar stores were gradually coming to an end. The power of the smartphone and the mobile Internet had just given them more options than ever.

Smartphone had turned into an ultimate a price-transparency device. While trolling stores, shoppers could compare prices and read reviews on similar products available from Amazon or the company's retail partners. That way the Seattle–based company encouraged smartphone-equipped consumers to link to its payments page and make a transaction. Customers who downloaded the app and enabled the location feature would also merit an additional 5 percent discount of up to $5 on Amazon's products. The service made it very easy for users to order the item via the app: With one phone tap, the often lower Amazon price for the item materialized in front of mobile user. A few more taps and that item was on the way to the mobile user's house.

APPS AND WEB GO HAND IN HAND

In the evolution of modern shopping, consumers had very recently progressed from visiting a physical store every time they wanted to buy something to shopping online via a desktop or laptop computer, at least some of the time. The next evolutionary step was to have shoppers make purchases online from mobile devices. Over the years, the smartphone intelligence had been steadily growing in parallel to the device's emerging status of being an important showcase of rich content with handset screens getting bigger and brighter.

It wasn't until 2008 that mobile commerce really began to grow, helped in large part by the advent of the iPhone, which generated enormous excitement for mobile commerce despite holding only a small chunk of the global smartphone market. Apps largely moved mobile commerce off the web and onto a more secure mobile Internet platform. They cut through the clutter of domain-name servers and uncalibrated information sources, taking users straight to the content they valued.

However, mobile users were also increasingly doing online browsing on their handsets. The game-changing iPhone had set the bar for a true mobile browsing experience in which sites rendered the way they would on a personal computer. Consumer expectations went up: they had good experiences with some mobile sites, and now they were expecting standard websites of other companies to offer a good mobile experience as well.

Mobile websites had an important role to play in the overall development of mobile commerce business. Case in point: while retailers pushed mobile users to download dedicated apps, many users were turning more and more to mobile websites. In mid-2010s, mobile users were spending more minutes on apps than mobile websites because native apps were more powerful than mobile websites. However, mobile websites would continue to get better through the maturation of HTML5, which could further narrow the gap between native apps and the mobile web.

Retailers embarking on e-commerce operations were increasingly building mobile-optimized sites because, as opposed to native apps, which worked on specific platforms like Android and iPhone, mobile sites rendered information to all Internet-enabled phones. Again, take the example of Uber, the cab ordering service which allowed city dwellers to quickly and easily get around through an app on their iPhone or Android devices. Once the service became popular, the San Francisco–based upstart launched a new mobile website m.uber.com to allow Blackberry and Windows Phone owners to use the service and thus expand the potential user base. Furthermore, mobile websites were inherently more suitable for specific user groups like stock market professionals. So, the power of apps aside, developing a mobile-friendly website would still be the foundation stone of a business' mobile game plan.

Mobile Internet was expected to surpass 2 billion users by 2015, thus outpacing desktop web usage. The real opportunity was in converting those browsers—who were growing by the day—into purchasers. If consumers spent more time browsing the web on their mobile devices than traditional devices like desktop PCs, they could ultimately end up shopping and purchasing more on those mobile devices. The opportunity clearly existed by making the mobile shopping experience as easy as possible for the consumer. A good mobile shopping experience was the one that was fully optimized for the smaller screen, took advantage of touchscreen technology, and offered a fast check-out in as few steps as possible.

Next up, the notion of the Internet of Things was closing in. Credit cards, for instance, were being equipped with smart chips so that these cards could communicate with the central nervous system otherwise known as the Internet of Things. That would eventually turn mobile commerce into an "Internet of Getting Things." So it was plausible that with the Internet of Things being chipped into the commerce landscape, mobile shopping would just become shopping and the rest would be the power of pervasive computing. No wonder Internet advertising giants like Facebook and Google were willing to spend billions to expand web services to those who didn't have it. They were apparently after the additional eyeballs and ad revenues that would come with the new user base.

MOBILE COMMERCE: A CASE STUDY | 223

  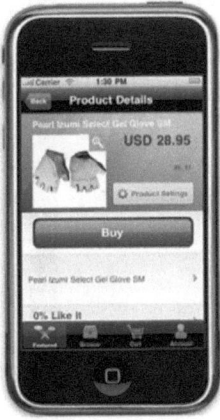

Mobile Internet—the primary vehicle of potentially the trillion dollar mobile commerce industry—provided a converged platform of location, payment and advertising systems through both the web and native apps.
Image source: Exadel Inc.

224 | **THE NEXT WEB OF 50 BILLION DEVICES**

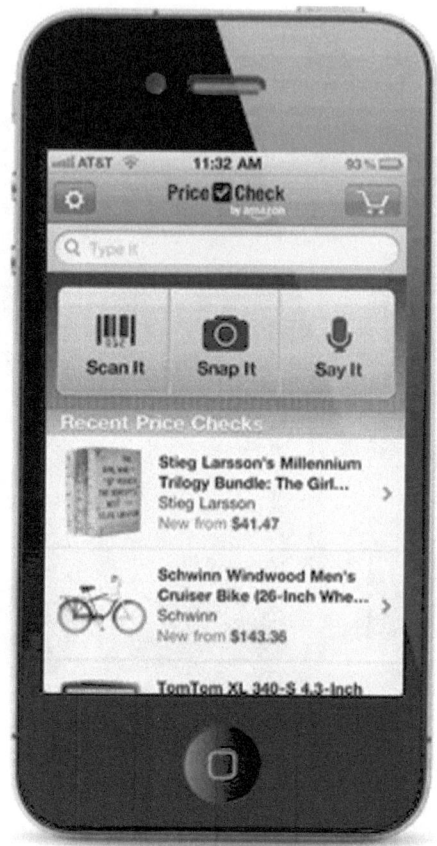

Amazon's Price Check mobile app led to the showrooming phenomenon and brought a huge disruption to physical retail operations. A mobile user could scan the product's arcode, snap a picture of it, speak its name, or just simply type its name. The app showed user prices from both online and offline retailers, including Amazon.
Image: Mashable

NOTES

Prologue

Majeed Ahmad, "Age of Mobile Data: The Wireless Journey To All Data 4G Networks," *CreateSpace*, March 5, 2014.

Chapter 1

"A Brief History of the Internet of Things," *Postscapes.com*

Cliff Edwards, "No Cartwheels for Handspring," *Bloomberg Businessweek*, April 2, 2001.

"Co-Creation and the New Web Of Things," *Digital Tonto*, April 11, 2012.

"History of M2M communications," *Wikipedia*.

Janna Anderson and Lee Rainie, "Digital Life in 2025," *The Pew Research Center*, March 11, 2014.

Majeed Ahmad, "Smartphone: Mobile Revolution at the Crossroads of Communication, Computing and Consumer Electronics," *CreateSpace*, December 16, 2011.

"PDA to Smartphone Evolution," *Techsplosive*, February 11, 2009.

Scott Moritz, "Verizon Pitches Its Wireless Network to Remote Oil Drillers," *Bloomberg Businessweek*, January 16, 2014.

Stacey Higginbotham, "Enjoying the internet of things? Thank your smartphone." *Gigaom*, November 26, 2012.

Tim Lohman, "The business benefits of machine to machine," *ZDNet*, January 9, 2013.

Chapter 2

"A mobile future," *The Economist*, October 11, 2001.

Andy Reinhardt, "Who Needs 3G Anyway," *Bloomberg Businessweek*, March 26, 2001.

Barnaby Page, "After Party, Europe's Telecoms Face Hangover," *TechWeb*, December 27, 2000.

Irene M. Kunii, "I-Way Bumps," *Bloomberg Businessweek*, May 15, 2000.

Irene M. Kunii, "Telecom Tremors," *Bloomberg Businessweek*, October 16, 2000.

Loring Wirbel, "Don't expect a false start," *EE Times*, December 19, 2000.

Loring Wirbel, "The road ahead," *EE Times*, September 27, 2000.

"Peering round the corner," *The Economist*, October 11, 2001.

Peter Elstrom, "Does Galvin Have The Right Stuff," *Bloomberg Businessweek*, March 17, 1997.

Rana Faroohar, "The Other Bubble," *Newsweek*, May 28, 2001.

Richard Comerford, "Handhelds duke it out for the Internet," *IEEE Spectrum*, August 2000.

Rick Merritt, "Software technologist sees cellular shakup coming," *EE Times*, September 26, 2002.

"Snap happy," *The Economist*, April 25, 2002.

"Son of Netscape," *The Economist*, August 10, 2000.

Stephen Backer and Kerry Capell, "Europe's Wireless Auctions: Give The Money Back," *Bloomberg Businessweek*, February 12, 2001.

"The Internet, untethered," *The Economist*, October 11, 2001.

"The wireless gamble," *The Economist*, October 14, 2000.

William Sweet, "Cell phones answer Internet's call," *IEEE Spectrum*, August 2000.

Yoshiko Hara, "Tomihisa Kamada—Tying home electronics to the Internet," *EE Times*, September 26, 2000.

Chapter 3

"A Finnish fable," *The Economist*, October 14, 2000.

Bolaji Ojo, "Consumer electronics at Apple's core," *EE Times*, July 9, 2007.

Claudine Beaumont, "RIM CEO tells Apple: 'You don't need an app for the web'" *The Telegraph*, November 17, 2010.

Dan Butcher, "Smartphone apps: the future of mobile advertising," *Mobile Marketer*, July 14, 2009.

David Rowan, "Mobile Makes Million—But It's Not as Simple as It Seems," *Wired*, April 10, 2011.

David Sarno, "BlackBerry vs. iPhone: What's in your pocket?" *Los Angeles Times*, December 5, 2010.

"Flash in the pan," *The Economist*, April 16, 2010.

Fred Vogelstein, "Behold, the Next Media Titans: Apple, Google, Facebook, Amazon," *Wired*, October 25, 2010.

Jane Black, "Will Investors Spring for Handspring?" *Bloomberg Businessweek*, March 14, 2002.

Jon Brodkin, "Google CEO Eric Schmidt: Smartphones will outsell PCs in two years," *Network World*, September 28, 2010.

Michael V. Copeland and Seth Weintraub, "Google: The search party is over," *Fortune*, July 29, 2010.

Richard Waters and Joseph Menn, "Space invader," *Financial Times*, June 6, 2010.

Rick Merritt, "Beyond the iPhone," *EE Times*, August 11, 2008.

Rick Merritt, "Opinion: Mobile Industry will eat Nokia's lunch," *EE Times*, September 1, 2009.

Seth Weintraub, "Larry Page: Jobs is rewriting history," *Fortune*, July 9, 2010.

Seth Weintraub, "Android less about money, more about iPhone disruption," *Fortune*, August 17, 2010.

"Seven mass media," *Wikipedia*, January 20, 2011.

"Understanding Smartphone Market Share? Battle not for phones, is for platform!" Tomi Ahonen's blog, July 30, 2010.

Chapter 4

Austin Carr, "Adobe CTO on MacBook Air, HTML5," *Fast Company*, November 8, 2010.

Brian X. Chen, "Will the Mobile Web Kill Off the App Store," *Wired*, December 18, 2009.

Chris Anderson and Michael Wolff, "The Web Is Dead. Long Live the Internet," *Wired*, August 17, 2010.

Dan Rowinski, "5 Trends In HTML5 In 2012," *ReadWrite*, December 26th, 2012.

Dan Rowinski, "Defining the Post-App Economy," *ReadWrite*, April 25, 2012.

Dan Rowinski, "Native Apps Versus Mobile Web: A Primer For Publishers," *ReadWrite*, October 8, 2012.

JP Mangalindan, "HTML5: not ready for primetime, but getting very close," *Fortune*, December 3, 2010.

JP Mangalindan, "Why rivals Google and Apple agree on HTML5?" *Fortune*, December 6, 2010.

Mathew Ingram, "Apps vs. the web: Are they enemies or allies?" *Gigaom*, December 14, 2011.

Mathew Ingram, "The rise of mobile apps and the decline of the open web — a threat or an over-reaction?" *Gigaom*, April 8, 2014.

Matt Baxter-Reynolds, "The mobile web is dead, long live the app," *ZDNet*, January 6, 2014.

Matt Silverman, "The History of HTML5," *Mashable*, July 17, 2012.

Nick Heath, "Web apps: the future of the internet, or an impossible dream?" *ZDNet*, August 16, 2013.

Nicholas Carlson, "This Engineer Saved Facebook," *Business Insider*, January 30, 2013.

"To HTML5 or not to HTML5, that is the mobile question," *Web Design Depot*, November 27, 2012.

Zachary Rosen, "In the era of mobile, the web will live on," *Fortune*, April 9, 2014.

Chapter 5

Brian S Hall, "The Next Billion Smartphone Users Will Have An Awesome Mobile Computing Device," *ReadWrite*, July 3, 2013.

Christopher Mims, "The five most disruptive technologies of 2012," *Quartz*, December 13, 2012.

Claire Cain Miller, "Browser Wars Flare Again, on Little Screens," *The New York Times*, December 9, 2012.

Dan Rowinski, "With Firefox OS, Mozilla Takes On The 'Closed' Internet—Again," *ReadWrite*, July 01, 2013.

David Meyer, "Why Opera's lightweight Mini browser is more popular than ever," *Gigaom*, January 24, 2013.

"History of Opera Software," *Wikipedia*.

Jessica Leber, "Big Mobile Data Bills: Could This Be the Makeover We Need?" *MIT Technology Review*, February 26, 2013.

JP Mangalindan, "The browser wars are back!" *Fortune*, June 4, 2012.

Rachel Metz, "The Browser Wars Go Mobile," *MIT Technology Review*, February 7, 2013.

Chapter 6

Bill Wasik, "Why Wearable Tech Will Be as Big as the Smartphone," *Wired*, December 17, 2013.

Brian Proffitt, "The Internet of Things might try to kill you," *ReadWrite*, September 18, 2013.

Clive Thompson, "Googling Yourself Takes on a Whole New Meaning," *The New York Times*, August 30, 2013.

Gene Marks, "How Google Screwed Up Google Glass," *Forbes*, April 21, 2014.

Heather Kelly, "Smartphones are fading. Wearables are next," *Fortune*, March 19, 2014.

Kevin McCullagh, "Why Wearable Devices Will Never Be As Disruptive As Smartphones," *Fast Company*, January 21, 2014.

Lee Rainie and Janna Anderson, "The Future of the Internet III," *Pew Research Internet Project*, December 14, 2008.

Marcelo Ballve, "Wearable Gadgets Are Still Not Getting The Attention They Deserve — Here's Why They Will Create A Massive New Market," *Business Insider*, August 29, 2013.

Chapter 7

Ben Popper, "Google's balloons versus Facebook's drones: the dogfight to send internet from the sky," *The Verge*, March 7, 2014.

Christopher Ryan, "The Next iPhone: Are We Ready for 4G?" *Gigaom*, March 22, 2010.

Glenn Fleishman, "The state of 4G: it's all about congestion, not speed," *Ars Technica*, March 29, 2010.

Leo Mirani, "These are the people who will build Facebook's drones," *Quartz*, March 28, 2014.

Lou Frenzel, "The Evolution Of LTE," *Electronic Design*, January 8, 2013.

Majeed Ahmad, "Age of Mobile Data: The Wireless Journey To All Data 4G Networks," *CreateSpace*, March 5, 2014.

PriyaGanapati, "Everything You Need to Know About 4G Wireless," *Wired*, June 4, 2010.

Steven Levy, "The Untold Story of Google's Quest to Bring the Internet Everywhere—By Balloon," *Wired*, August 13, 2013.

Chapter 8

Arik Hesseldahl, "Apple's iDecade," *Bloomberg Businessweek*, April 26, 2010.

Brian Proffitt, "The IoT in 2014: Steady as it goes," *ReadWrite*, December 27, 2013.

Jesse Berst, "Don't Be Seduced By the M-Commerce Siren Song," *ZDNet*, April 17, 2000.

Junko Yoshida, "Nokia, not Google, sees itself reshaping the Internet," *EE Times*, February 11, 2008.

Junko Yoshida, "Nokia's naked ambition: Moving beyond cellphones," *EE Times*, October 23, 2008.

Majeed Ahmad, "Mobile Commerce 2.0: Where Payments, Location and Advertising Converge," *CreateSpace*, October 15, 2013.

"Net-enabled cell phones create m-commerce," *Tribune*, October 19, 2000.

INDEX

2.5G, 184-185
3G, third-generation wireless, 16, 84, 186-190
3GDoctor, 21
4G, fourth-generation wireless, 1, 7, 26, 27, 191-194, 196
5G, fifth-generation wireless, 1, 26
accelerometer, 20
Access Co. Ltd, 40
Amazon, 112-113, 116, 205-206
Andreessen, Marc, 1, 136
Android, 4, 18, 71-72, 128, 138-139
Android Nation, 138
Android Wear, 154-155
AOL, America Online, 76, 77, 125, 210
app Internet, 96-101
Apple, 3-4, 12-13, 19, 20, 61, 68-71, 72-74, 77-78, 85-87, 91-92, 94, 99, 104-105, 107, 108-109, 113, 202

App Store, 106, 108, 109, 113, 202
Ascenta, 174-175
Asha phone, 138
Ashton, Kevin, 14-15, 30
AT&T, 24, 25, 45, 186
augmented reality, 165, 166, 172
balloons, 176-179
BlackBerry, 76, 78-79, 96, 185
Bluetooth, 143, 150, 153, 200
Boot2Gecko, 122
Bush, Vannevar, 151, 171
Chambers, John, 157, 168
Chrome-lite, 95
cHTML, 40, 50
Cisco System, 15, 157, 168-169, 191
City Lens, 164
cloud computing, 3, 129, 130, 139-143, 164
context-aware Internet, 199-200
Danger Inc., 77
Dolphin, 127
drones, 174-176, 179-180, 197

Eich, Brendan, 124-125, 144
Enoki, Keiichi, 37-38, 63
Ericsson, 7, 33, 45, 46, 174, 183
Facebook, 3, 106-108, 162, 163, 173-176, 179, 197
Firefox, 101, 122
Firefox OS, 121-124, 135, 138
FitBit, 160
Flash, Adobe Flash, 99, 101-103
Flurry, 111
Foursquare, 164-165
Futureful, 126-127
Galvin, Christopher, 42
Gershenfeld, Neil, 15
GM, General Motors, 24
Google, 19, 71-74, 85, 96-97, 99, 103-104, 138, 143, 151, 164, 176
Google Fiber, 169
Google Glass, 28, 143, 149-153, 156, 172
Google Now, 149-150
Google Play, 106, 108
GPRS, General Packet radio Service, 182-183
GPS, global positioning system, 19, 23, 152, 164, 166, 200, 201
GSM, Global System for Mobile Communications, 36, 182, 187
Handspring Treo, 77
Hickson, Ian, 120
Hiptop, 77
HitchHiker, 127
HTML5, 99-100, 102-112, 114-118, 120, 121-125, 129-130, 141
hybrid HTML5 apps, 116
i-mode, 37-41, 49-51, 55, 61, 68, 210
Industrial Internet, 9
Internet of Everything, 157, 167-170
Internet of Things, IoT, 9-10, 14-18, 24-29, 142, 157-161, 167, 222
Internet.org, 173-174
iOS, 69, 113
iPhone, 3-5, 18-23, 67-70, 77-78, 84, 89-90, 91-92, 159, 185-186, 189, 201, 202-203, 215-218
ITU, International Telecommunications Union, 2, 15, 192

INDEX

iTunes, 85, 86, 93
Ivarsøy, Geir, 130-131, 147
jailbreaking, 91
Japan, 38-39, 41-44, 50
Java, 101, 102, 125, 144
JavaScript, 109-110, 125, 144
Jawbone UP, 160
Jobs, Steve, 13, 67, 81, 85, 89, 90, 91, 101-102, 119
Layar, 166
Lee, Steve, 151-152
Levinson, Arthur, 119
LG Electronics, 15, 18, 160
LiveSight, 165
Logica plc, 40
LTE, Long Term Evolution, 1, 7, 8, 9, 24, 26, 29, 192-194, 196
M2M, machine-to-machine communications, 8-10, 16-18, 25
Matsunaga, Mari, 65
MediaTek, 138-139
Meeker, Mary, 72, 134, 162
MEMS, micro electro-mechanical system, 21, 22
Microsoft, 75, 129-130, 137, 161
MIT, Massachusetts Institute of Technology, 14, 15
MMS, multimedia messaging service, 55-57
mobile apps, native apps, 83, 91-101, 104, 112-118, 119, 202, 217
mobile browser, 40, 45, 47, 61, 95, 101-106, 118, 125-134, 137
mobile commerce, 181-182, 195, 204-224
mobile Internet, 1-5, 19, 27, 33-65, 67-90, 95, 134-141, 182, 189, 190, 196
Mobile Internet Explorer, 95, 127, 130
mobile web, 91-92, 95, 112-118, 220-221
MoboTap, 127
Morgan Stanley, 1, 80
Motorola, 42, 45, 46
Mozilla, 101, 121-125, 135, 138
Mozilla Foundation, 99, 125
MultiTorg Opera, 131
Natsuno, Takeshi, 41, 64
Navteq, 199

Nest, 160
NetHopper, 127
Netscape, 125, 144, 129, 187
Netscape Navigator, 51, 125, 144
Newton MessagePad, 3, 12-13, 18, 145
Nokia, 19, 33, 45, 46, 75, 79-80, 136-138, 165, 199-201
Nokia Nearby, 137-138
NTT DoCoMo, 37-44, 57, 68, 82, 210
Ohboshi, Koyji, 37
Ollila, Jorma, 58, 80
Ondrejka, Cory, 107
OnStar, 24
Opera Mini, 128, 130, 131, 132, 133-134, 146
Opera Mobile, 133-134
Opera Software ASA, 99, 128, 130-134, 147
Orr, Cheryl Langdon, 135
PacketNet, 45
Palm, 78
Palm Pilot, 13
Palm Treo, 77
PARC, Xerox Palo Alto Research Center, 10, 11, 27, 59
Parviz, Babak, 151
PDA, personal digital assistant, 3, 12-13, 76
Phone.com, 46
PHS, Personal Handyphone Service, 43
PocketWeb, 127, 145
Price Check, 113, 218-219, 224
Project Loon, 176-179
Qualcomm, 25
responsive design, 28-29, 115
RFID, 13-15, 30, 150, 161
Rockmelt, 126
Rossman, Alain, 45, 62
Safari mobile browser, 4, 69, 91, 95, 109, 128
Schmidt, Eric, 90
Scully, John, 12-13
sensors, 20-23, 158-159
Sidekick, 77
Siemens, 16-17
Silverlight, 99
Siri, 163-164
Skyfire, 132-133
smart home, 25-26
smartphone, 22, 27, 28, 81, 87, 140, 142-143, 151, 155-156
smartwatch, 153-154

SMS, short message service, 35, 54, 58, 162-163, 215
SPOT, smart personal objects technology, 161
Starner, Thad, 152, 172
StatCounter, 136
STNC, 127
Symbian, 75, 137
TecO, 127, 145
Teledesic, 177
Telenor, 131, 147
Televerket, 131
Tetzchner, Jon Stephenson von, 130-131, 147
thermostat, 160
Titan Aerospace, 179
Twisthink, 8
Uber, 114, 221
ubiquitous computing, 11, 19, 27
UIWebView, 109
Unwired Planet, 45-46
Verizon, 7-8, 25, 169
W3C, World Wide Web Consortium, 99-100, 108, 120
WAP, Wireless Application Protocol, 45-52, 55, 57, 59, 61, 69, 82, 200, 210, 213-214
wearable computing, wearable devices, 28, 142-143, 149-161, 171
Web APIs, 122-123
web apps, 122-123, 137-138
Web Pass, 132-133
Weiser, Mark, 10-11, 18, 27, 30
WhatsApp, 162-163
WHATWG, Web Hypertext Applications Technology Working Group, 99-100, 120
Whitacre, Ed, 194
WML, wireless markup language, 47, 48
Xpress browser, 137
Yahoo!, 76, 126
Zuckerberg, Mark, 106, 107, 173-174, 180

ABOUT THE AUTHOR

Majeed Ahmad is former Editor-in-Chief of *EE Times Asia*, a sister publication of *EE Times*. While being the Editor-in-Chief at Global Sources, a Hong Kong–based publishing house, he also spearheaded magazines relating to electronic components, consumer electronics, and computer, security and telecom products.

This is his sixth book on wireless and smartphones. His first five book titles are *Smartphone*, *Nokia's Smartphone Problem*, *Mobile Commerce 2.0*, *Age of Mobile Data*, and *Essential 4G Guide*.

He is currently associated with a number of technology publications as a contributing writer and Editor-at-Large. He has been a technology and trade journalist for more than 18 years.